Task Analysis

EDITED BY

John Annett

Department of Psychology, University of Warwick, UK

AND

Neville A. Stanton

Department of Design, Brunel University, UK

London and New York

First published 2000 by Taylor & Francis
11 New Fetter Lane, London EC4P 4EE

Simultaneously published in the USA and Canada
by Taylor & Francis Inc,
29 West 35th Street, New York, NY 10001

Taylor & Francis is an imprint of the Taylor & Francis Group

Typeset in Times
Printed and bound in Great Britain by T. J. International, Padstow, Cornwall

Every effort has been made to ensure that the advice and information in this
book is true and accurate at the time of going to press. However, neither the
publisher nor the authors can accept any legal responsibility or liability for any
errors or omissions that may be made. In the case of drug administration, any
medical procedure or the use of technical equipment mentioned within this book,
you are strongly advised to consult the manufacturer's guidelines.

British Library Cataloguing in Publication Data
A catalogue record for this book is available from the British Library

Library of Congress Cataloguing in Publication Data
Annett, John.
Task analysis / John Annett and Neville Stanton.
p.   cm.
This book is a production of a 1998 issue of the journal Ergonomics,
with a new introd. by the authors
Simultaneously published in the USA and Canada.
Includes bibliographical references and index.
Romanized record.
1. Human Engineering. 2. Task analysis. 3. Cognitive science.
I. Stanton, Neville, 1960– . II. Title.
TA166.A47 2000
620.8′2–dc21   99-044790

ISBN 0-7484-0906-8

Contents

CHAPTER 1

Research and developments in task analysis

JOHN ANNETT and NEVILLE STANTON

Department of Psychology, University of Warwick, Coventry CV4 7AL, UK
Department of Psychology, University of Southampton, Highfield, Southampton
SO17 1BJ, UK

1.   Introduction

Methods of collecting, classifying and interpreting data on human performance in
work situations lie at the very root of ergonomics. Task analysis, as these methods are
collectively known, reflects both our current understanding of factors affecting human
performance and the information needs of system designers. Modern technology has
changed the nature of human work dramatically in recent years and concepts and
techniques of analysis have changed and are changing to meet new needs. This book,
based on a special issue of *Ergonomics*, provides a timely opportunity to review
current theory and practice and to identify future trends.

2.   Origins

The 'Scientific Management' movement at the turn of the century (Gilbreth 1911;
Taylor 1911) was primarily concerned with the analysis of physical work that could
lead to the economical design of work places and methods of production. Technological developments during the Second World War refocused attention on the
information processing aspect of tasks and the role of the human operator as a
controller, planner, diagnostician and problem solver in complex systems. Such
systems are prone to catastrophic failure, sometimes attributable to human error, and
so new concepts of the limits of human performance and methods of analysis were
developed (Chapanis 1959; Craik 1947, 1948; Fitts 1951). The process flow chart of the
earlier period was adapted to the analysis of communication links (Chapanis 1959)
and the analysis of statistical data on aircraft accidents (Fitts and Jones 1947 a, b) was
developed into the Critical Incident Technique (Flanagan 1954).

As tasks became more demanding of specialized skills and knowledge the cognitive
components of human performance began to form a special focus of interest in relation
to training requirements, and this was enhanced by the development of education and
training technology during the 1950s and 1960s (Glaser 1962; Miller 1962). During
this period the principle of hierarchical decomposition (Annett *et al.* 1971) was
introduced to replace the classic linear process chart as a means of capturing the
psychological complexity of tasks that involved significant cognitive activity such as
planning, diagnosis and decision making. It was also during this period that
psychologists (for it was mostly psychologists who were developing these methods)
began to abandon the classical behaviourist concepts of stimulus and response and to
talk of cues, decisions, feedback and knowledge in talking of human performance.
Thus the foundations for a group of methods which have come to be known
collectively as Cognitive Task Analysis (CTA) were laid. These methods have been
greatly influenced by the technical challenge of Human-Computer Interaction (HCI)
and the modelling of human performance and the use of formal methods derived from

software engineering (Card *et al.* 1983). This developing methodology (CTA) is well represented in this volume.

3. Hierarchical task analysis

Hierarchical Task Analysis (HTA) developed by Annett *et al.* (1971) is probably the most frequently used methodology by ergonomists in the UK and is often incorporated into CTA, so it makes a useful point of reference when making comparisons with other approaches. HTA is perhaps popular primarily because of its generic nature, it may be used to analyse any type of task in any domain. The key feature of HTA is that *tasks*—those things that the person is seeking to achieve—are essentially defined by *goals* rather than actions as such, and that complex tasks may be analysed by decomposing a hierarchy of goals and sub-goals.

HTA was influenced by a control theory model of human performance embodied in the idea of Test-Operate-Test-Exit or TOTE units (Miller *et al.* 1960). The essence of a TOTE unit is that it specifies a goal, represented by the set point of a control loop, and an action generation system controlled by feedback on the discrepancy between set point and current goal state. A nested hierachy control unit ('operations' in HTA terminology) is the essential feature of the analysis since goals can have many sub-goals and there is a parent-child relationship at each level in the hierarchy. The level of detail to which an analysis is taken is not fixed, as in both classical work study and recent HCI methods, but depends on whether or not the source of performance failure (or significant risk of failure) can be identified. HTA has been linked to many applications including personnel selection and manning, training, workplace layout and equipment design and this versatility is perhaps the main reason behind its success.

Many investigators have taken HTA as the basis for development of their own methods. Astley and Stammers (1987), for example, used HTA as the basis for investigating information flows between people and computers; Bass *et al.* (1995) developed a computer-based, object-oriented, version of HTA; Embrey (1986) developed a HAZOP-type human error prediction methodology; Baber and Stanton (1994) developed a scenario-based analysis technique and Annett (1996, 1997; Annett and Cunningham, in press) have developed HTA for analysis of team tasks.

4. New developments

However, critics of HTA are quick to identify potential weaknesses. It is argued that there is a craft-skill requirement to conducting a sensitive HTA, which is due to the subjectivity in interpretation of data and ambiguity in the analytic process. It has also been suggested that there are inconsistencies in the way in which HTA has been used by different practitioners. Advocates of HTA claim that this is missing the point of the methodology, since the purpose of HTA is to focus on the analysis of the task, to understand what people are doing, how they can fail, and how such failures might be prevented. Rather like soft-systems methodologies (e.g. Checkland 1990) the process of doing HTA is worth rather more than the final representation on paper because the analyst now understands the problem. When properly applied, HTA provides both a diagnosis and a proposed cure, not just a list of symptoms.

Nevertheless, Stammers (1995) warns that further developments in task analysis techniques in general will be held back by the lack of standardization of terminology, methodological diversity and limitations in transferability. The proliferation of HTA-style techniques may be symptomatic of these problems. As Wilson (1995) observes:

' . . . method which to one researcher or practitioner is an invaluable aid to all their work may to another be vague or insubstantial in concept, difficult to use and variable in its outcome' (Wilson 1995: 21).

The chapters in this volume illustrate a number of still unresolved issues and areas of concern.

(1) First there are issues concerning the theoretical orientation and justification of the concepts and methods used. What are the assumptions underlying accounts of human performance and the logical status of the notations used to express them?

(2) Next there is range of practical problems and methodological questions such as how to collect data and how to present and draw conclusions from the results. Here questions of practicality are clearly important but questions of validity and reliability of the various methodologies still need to be addressed.

(3) Third we have different fields of application. All the papers concern tasks of some complexity but training, system design, collaborative working and error prediction emerge as highly significant areas of application.

5.   Contents of this volume

All the papers have something to say about each of these issues. The authors exhibit a wide range of theoretical backgrounds and assumptions ranging from classical human factors/ergonomics through cognitive psychology, software engineering and systems design. A notable distinction is between classical approaches and those informed by relatively more recent developments in cognitive theory and software engineering. Most of the papers have a 'cognitive' orientation, which is hardly surprising in view of the effects of automation and computerization in reducing the demand for routine repetitive manual work and in providing a wealth of complex tasks which depend on planning, decision making and problem solving. Any methodology that cannot cope with the mental aspects of a task is hardly worth considering.

Shepherd (chapter 2) makes a spirited defence of Hierarchical Task Analysis that, in a way formed a historical and conceptual bridge between the classical and cognitive viewpoints, and is today probably the most commonly used approach. Shepherd claims that the decomposition of goals, goal hierarchies and plans can provide a general framework for all task analysis and not just a another toolkit. He goes on to compare the 1960s and 1970s control theory concepts of TOTEs and Operations with the GOMs (Goals, Operations, Methods) and Production Systems of the 1980s and concludes the differences are not fundamental and indeed that 'Operations' permit different levels of description or degrees of 'graininess' not recognized in these later formulations.

A more subtle issue is whether task analysis should be viewed simply as a methodology for describing the actual behaviour of job-incumbents or whether it should be aimed at producing a more theoretical analysis of the demand characteristics of the task, or perhaps a model of the cognitive processes that must occur inside the head of anyone performing the task. Modelling is of great potential value in the design of complex socio-technical systems such as air traffic control, military command and control and chemical and nuclear process control systems that require thorough evaluation prior to commissioning. Most analyses start by attempting to discover what people actually do even if, as in Marti's chapter on structured task analysis in complex domains (chapter 8), the aim is to identify task requirements and to reallocate functions between live operators and supporting technology in future systems still on the drawing board.

Coming from a system design perspective Diaper, McKearney and Hurne (chapter 3) are critical of current task analysis techniques for not being compatible with software engineering methods. They argue that ergonomics must develop methods that can be used by software engineers rather than expecting software engineers to adapt to Ergonomics methods. They propose Data Flow Analysis as a central technique for software engineering as it represents the data flow between elements in the computer system. Their Pentanalysis technique was designed to capture the transactions between people, people and artefacts, and between artefacts, as well as the content of those transactions. As the name implies, the Pentanalysis comprises five main parts that integrate data from task analysis and data flow diagrams. If the technique leads software engineers to design more human-centred systems, then Diaper *et al.* will have succeeded in encouraging ergonomics to take a more central role.

Watts and Monk (chapter 4) argue that decompositional techniques are not very successful in representing synchronization and collaborative activity (but see Swezey *et al.*, chapter 9). They criticize contemporary methods for representing computer-supported collaborative activity for being too restrictive and forcing the analyst to adopt an inappropriate framework. Drawing upon experience of case study material they develop a scenario-based analysis method called the Communication technologies Usage Diagram (CUD). The case study of medical teleconsultation shows the use of combined audio and video in which verbal, non-verbal gestures and visual cues are manifest during the course of the consultation. The combined media provide a richer interaction than would be provided with either alone. The CUD shows the activities and communication links between different sites together with the communication resources and the effects of communication technologies on all participants. Watts and Monk claim that it is not necessary to represent the fine grain sequence of events as this overly complicates the analysis, and that it is sufficient to identify the phases that are passed through. This level of detail is sacrificed in order to improve the overview and readability of the analysis.

With regard to methodological issues Ainsworth and Marshall (chapter 5) believe that task analysis must be capable of addressing many different questions on a wide variety of different systems and that little is known of the validity and reliability of different methods. They describe two surveys involving 90 analyses of tasks in the defence and nuclear industries which are highly cognitive and characterized by time criticality of decision making. While HTA was found to be 'the nearest thing to a universal TA technique' HTA plans were only developed in a minority of cases and both the quality and depth, as judged by a small group of experts, was often questionable. Analyses conducted by the nuclear industry show much more consistency than those in the defence industry and are more often carried out by human factors specialists. TA is often introduced earlier in design process in nuclear than in defence application but methods of obtaining data in all contexts are very variable. Systematic decomposition of goal hierarchies is not universal and stopping rules of not always applied. The surveys indicate that there is a need for more published guidelines and better training of analysts.

Millitello and Hutton (chapter 6) recognize the methodological problems facing relatively untrained individuals and in Applied Cognitive Task Analysis (ACTA) have developed a streamlined methodology with explicit procedural guidelines. In a series of structured interviews with subject matter experts (SMEs) various kinds of information are gathered. A Task Diagram describes the overall structure of the task and highlights cognitive difficulties; a knowledge audit identifies the expertise required to perform the task and a simulation interview provides an operational context by working through a specific scenario. Finally the analyst draws up a table of cognitive demands that helps to identify strategies and common errors. Militello and Hutton address the problems of learnability and usability of ACTA by reporting a study in which students were taught the method in six hours of seminars and then applied the method by interviewing SMEs in two specialist areas. Usability was found to be good, given the well-structured ACTA approach but more work is needed to establish meaningful metrics for reliability and validity.

Omerod, Richardson and Shepherd (chapter 7) are also concerned with making HTA more usable but they argue that the use of notation that can make methods more consistent may actually hamper the analysis process. By way of a solution, they offer the Sub-goal Template (SGT) method, in which a set of standard task elements may be used to classify tasks into action, communication, monitoring and decision types and sequences of tasks to be classified into fixed, contingent, parallel and free types. The SGT notation and method can be incorporated within HTA to make the process of data collection more efficient and in this way the SGT approach can give the HTA process firmer direction. These authors developed the method through empirical trials by analysing the errors made by novices. The results led to the development of a computer-based approach to SGT, which guides users through the steps of the analysis. In general they found that computer-assisted participants solved more problems and made fewer errors than their counterparts using only printed instructions. However, the computer-based group also took longer to complete the task initially, but did speed up with practice over six trials. This study is particularly encouraging for the provision of computer support for task analysis.

The breadth of applications of task analysis is illustrated by the inclusion of papers concerned with system design, team training and error identification.

Marti's chapter (chapter 8) is concerned with modelling air traffic control tasks for design purposes. The analysis uses task graphs to represent the goal structure with task/sub-task decomposition, logical constraints, triggering conditions, loops, etc. and also represents objects and multi-user roles. The elements of the notation for the General Task Model (GTM) have much in common with HTA. A key problem is how to proceed in a principled way from the general task model to a composite task model in which alternative configurations by which tasks are re-allocated between humans and computers can be considered. This key transition still remains largely a matter of intuition.

Air Traffic Control tasks require teamwork, communication and co-ordination. Swezey, Owens, Bergondy and Salas (chapter 8) on Task and Training Requirements Analysis Methodology (TTRAM) note that 'traditional task analysis techniques do not provide mechanisms for capturing requisite cues, conditions and standards for team tasks and skills'. TTRAM is based on obtaining estimates, that is ratings, by SMEs of various features of tasks that will impinge on training requirements (specifically teamwork, skill decay, practice opportunities) and goes on to estimate the potential of simulation to meet these requirements. Appropriate training technologies are determined by first considering each task's skill decay index as compared to its practice effectiveness. In addition important skills underlying performance, the task's criticality and internal and external teamwork requirements are considered in the methodology. The paper provides an excellent illustration of the sometime forgotten precept that task analysis should be a way of providing solutions, not just of describing problems.

O'Hare, Wiggins, Williams and Wong (chapter 10) define cognitive task analysis after Redding (1992) as a method of determining 'the mental processes and skills required to perform a task at high proficiency levels and the changes that occur as skills develop'. Their approach draws on Klein's Recognition-Primed Decision Model (RPDM) and Critical Decision-Making Method (CDM), which itself is descended from the classical Critical Incident Technique. The method is applied to developing training media in the fields of aviation, white-water rafting and emergency control. Descriptions of critical incidents are expanded by using 'cognitive probes' specifically designed to allow the SME to elaborate on situational awareness and planning skills which are central to time-pressured decision-making skills.

Shryane, Westerman, Crawshaw, Hockey and Sauer (chapter 11) consider the application of human error assessment techniques to the design of safety-critical software in a railway signalling control system. They chose HTA as the starting point for the analysis. Data were initially collected from interviews, existing documentation and observation methods and combined into the HTA. As a description of normative behaviour, the HTA focused attention on possible sources of error. Four main methods were used to provide insight into potential errors: error logs, observation, work samples and laboratory investigation. The authors argue that multiple sources are required, as each will incorporate some degree of bias although some degree of convergence would add additional support. They also argue that a matrix of methods combining a variety of empirical methods is likely to be superior to that of expert judgement. The empirical methods seemed to produce different kinds of data, although there was some compatibility between the observation and work sample methods and between the logs and task simulation exercise. The exact degree of convergence between the methods is difficult to judge. Shryane *et al.* point out the difficulty in representing the quantitative aspects of the task (i.e. frequency of performance) in the analysis and suggest that very low event frequencies make error analysis particularly difficult.

Stanton and Stevenage (chapter 12) consider a variety of Human Error Identification (HEI) techniques which may be evaluated in terms of their reliability and validity, their learnability and usability. They describe a study using an error identification technique, Systematic Human Error Reduction and Prediction Approach (SHERPA). Experimental subjects were taught the principles of HTA and the SHERPA methodology and were required to predict error rates on a vending machine task for which empirical error rates had been determined. The experimental group which had received training in both HTA and SHERPA made more accurate error predictions so once more we have an example of a method of analysis that produces not just descriptions but solutions. Perhaps the ultimate test of a task analysis method is that it can model performance with sufficient accuracy to make quantitative predictions about the performance of real operators performing the task. Both theory and methodology are currently some way short of this target but the gap between craft skill and applied science is narrowing as we improve assessment of the reliability and validity of analytical techniques.

6. Conclusions

In conclusion, we believe that the contributions to this book comprise a representative cross-section of contemporary task analysis research and raise important issues for future research and development. Among those issues that stand out from the present collection of papers are problems of the appropriateness to and learnability by different user groups, the range of application and reliability and validity of different task analytic methods.

Many chapters deal with the appropriateness of the task analysis techniques to particular user groups, and suggest ways in which shortcomings might be dealt with either by adding functionality, or by the provision of guidance and software support tools. Linked to developing analytical techniques to suit the specific user group is the problem of the learnability of the techniques themselves. Several contributions show that people without a formal training in ergonomics can acquire these skills and techniques given the appropriate training and support. Moreover, the performance of novices improves dramatically with practice.

All the authors are concerned with extending the capability of task analysis techniques to capture as rich a picture of human activity as may be desired, including specifying normative performance standards, identifying sources of error, the verbal and non-verbal knowledge required for performance, and means of referring to both individual and team performance. Significant developments have taken place in all these aspects, including enhancement of existing techniques and the development of new techniques.

A continuing feature of task analysis methodology is the need to integrate data from a variety of different sources in order to provide some degree of internal validation as well as a means of developing an information-rich picture of the task. The extent to which different sources of information may be relied upon for an accurate analysis of the task is an issue of continuing debate. While some results are encouraging, more studies of reliability and validity need to be conducted.

In conclusion, each of these four issues could provide a programme of research to take task analysis into its second century of development. We hope that in some small way we have helped stimulate the research community to carry the 'task analysis' torch forward into the next millennium and thank all of the contributors to this volume.

References

ANNETT, J. 1996, Recent developments in hierarchical task analysis, in S. A. Robertson (ed.), *Contemporary Ergonomics, 1996* (London: Taylor & Francis), 263–268.

ANNETT, J. 1997, Analysing team skills, in R. Flin, E. Salas, M. Strub and L. Martin (eds), *Decision Making Under Stress* (Aldershot: Ashgate), 315–325.

ANNETT, J. and CUNNINGHAM, D. (in press) Analysing command team skills, in J. M. C. Schraagen, S. F. Chipman and V. L. Shalin (eds), *Cognitive Task Analysis* (Mahwah, NJ: Erlbaum).

ANNETT, J., DUNCAN, K. D., STAMMERS, R. B. and Gray, M. J. (1971). *Task Analysis* (London: HMSO).

ASTLEY, J. A. and STAMMERS, R. B. 1987, Adapting hierarchical task anaysis for user-system interface design, in J. R. Wilson, E. N. Corlett and I. Manenica (eds), *New Methods in Applied Ergonomics* (London: Taylor & Francis), 175–184.

BABER, C. and STANTON, N. 1994, Task analysis for error identification: A methodology for designing 'error tolerant' consumer products, *Ergonomics*, **37**(11), 1923–1941.

BASS, A., ASPINAL, J., WALTER, G. and STANTON, N. A. 1995, A software tool for hierarchical task analysis, *Applied Ergonomics*, **26**(2), 147–151

CARD, S. K., MORAN, T. P. and NEWELL, A. 1983, *The Psychology of Human-Computer Interaction* (Hillsdale NJ: Erlbaum).

CHAPANIS, A. 1959, *Research Techniques in Human Engineering* (Baltimore: Johns Hopkins University Press).

CHECKLAND, P. 1990, *Soft Systems Methodology in Action* (Chichester: Wiley).

CRAIK, K. J. W. 1947, Theory of the human operaor in control systems. I: The operator as an engineering system, *British Journal of Psychology*, **38**, 56–61.

CRAIK, K. J. W. 1948, Theory of the human operator in control systems. II: Man as an element in a conrol system, *British Journal of Psychology*, **38**, 142–148.

EMBREY, D. E. 1986, SHERPA: A systematic human error reduction and prediction approach. Paper presented at the International Meeting on Advances in Nuclear Power Systems, Knoxville, Tennessee.

FITTS, P. M. 1951, Engineering psychology and equipment design, in S. S. Stevens (ed.), *Handbook of Experimental Psychology* (New York: John Wiley & Sons), 1287–1340.

FITTS, P .M. and JONES, R. E. 1947a, Analysis of factors contributing to 460 'Pilot Error' experiences in operating aircraft controls. Memorandum Report TSEAA-694-12, Aero-Medical Lab., Air Materiel Command, Wright-Patterson Air Force Base, Dayton, Ohio.

FITTS, P. M. and JONES, R. E. 1947b, Psychological aspects of instrument display. 1: Analysis of 270 'Pilot Error' experiences in reading and interpreting aircraft instruments. Memorandum Report TSEAA-694-12A, Aero-Medical Lab., Air Materiel Command, Wright-Patterson Air Force Base, Dayton, Ohio.

FLANAGAN, J. C. 1954, The critical incident technique, *Psychological Bulletin*, **51**, 327–358.

GILBRETH, F. B. 1911, *Motion Study* (Princeton NJ: Van Nostrand).

GLASER, R. (ed.) 1962, *Training Research and Education* (New York: Wiley).

MILLER, R. B. 1962, Task description and analysis, in R. M. Gagne (ed.), *Psychological Principles in System Development* (New York: Holt, Reinhart & Winston), 187–228.

MILLER, G. A., GALANTER, E. and PRIBRAM, K. 1960, *Plans and the Structure of Behavior* (New York: Holt).

REDDING, R. E. 1992, A standard procedure for conducting cognitive task analysis. Report to the Federal Aviation Administration, Human Technology Inc., Mclean, VA.

STAMMERS, R. B. 1995, Factors limiting the development of task analysis, *Ergonomics*, **38**(3), 588–594.

TAYLOR, F. W. 1911, *Principles of Scientific Management* (New York: Harper & Row).

WILSON, J. R. 1995, A framework and context for ergonomics methodology, in J. R. Wilson and E. N. Corlett (eds), *Evaluation of Human Work: A practical Ergonomics methodology*, 2nd edn. (London: Taylor & Francis).

CHAPTER 2

HTA as a framework for task analysis

A. SHEPHERD

Department of Human Sciences, Loughborough University, Loughborough
LE11 3TU, UK

Keywords: Task analysis; Cognitive task analysis; Complex tasks.

The motivation for this paper is to review the status of Hierarchical Task Analysis (HTA) as a general framework for examining tasks, including those for which cognitive task analysis methods might be assumed to be necessary. HTA is treated as a strategy for examining tasks, aimed at refining performance criteria, focusing on constituent skills, understanding task contexts and generating useful hypotheses for overcoming performance problems. A neutral and principled perspective avoids bias and enables the analyst to justify using different analytical methods and develop hypotheses as information is gained about the task. It is argued that these considerations are equally valid when examining tasks that are assumed to contain substantial cognitive elements. Moreover, examining cognition within the context of a broader task helps to situate cognition within the network of actions and decisions that it must support, as well as helping to establish where effort in cognitive task analysis is really justified.

1. Introduction

Hierarchical Task Analysis (HTA) is a term that encompasses ideas developed by Annett and Duncan (Annett and Duncan 1967, Annett *et al.* 1971, Duncan 1972, 1974). It explores tasks through a hierarchy of *goals* indicating what a person is expected to do, and *plans* indicating the conditions when subordinate goals should be carried out. Rather than setting out to establish a distinctive method of task analysis, the original work suggested general principles for guiding all task analysis projects. HTA has enjoyed substantial success in many practical applications. This is illustrated in Kirwan and Ainsworth (1992) who provide a number of case studies describing task analysis applications in which HTA features prominently. Despite its successes, it is often assumed that HTA is unsuited to dealing with cognitive tasks. This paper will review the status of HTA as a framework for conducting task analysis rather than as a method for modelling behaviour. In this respect the paper aims to demonstrate how even the analysis of cognitive tasks benefits from consideration of the wider task perspective that HTA offers.

Contributions to the task analysis literature have often described techniques rather than discuss fundamental issues. This has encouraged a 'tool-kit' approach where different methods are offered for use at the discretion of the analyst. A problem with this approach is that it relies on the analyst prejudging the nature of the task to be examined or the nature of the solution. This can lead to bias, error of judgement and unfocused work, unsuited to an applied setting. Identifying appropriate analytical methods should be part of task analysis and not a precursor to it. Thus Duncan (1972) argues that task analysis should be neutral with respect to solution. Viewing HTA as a distinct technique has tended to place it in the tool-kit along with other methods that appear to do specific things, such as *link analysis* or *time-line analysis* or, indeed, methods of *cognitive task analysis*. Selecting task analysis methods should be governed by a principled examination of task demands and constraints identified through the course of the task analysis project. It will be argued that HTA provides such a framework of principles.

Debate about task analysis is hampered by a lack of agreement of the nature and purpose of task analysis. The word 'task' may be treated as a reference to human behaviour, the system goals for which people are employed, how context constrains the attainment of goals, or some interaction between these and other factors. 'Task analysis' may be treated as an investigatory tool or a method of modelling human behaviour. Task analysis may be seen as a specific and rigorous method or merely a guiding framework. 'Task' is seldom defined satisfactorily. One of the few definitions of 'task' is Miller's (Miller *et al.* 1953:1): a task 'is a group of discriminations, decisions and effector activities related to each other by temporal proximity, immediate purpose and a common man-machine output'. This definition combines features of human behaviour and features of systems behaviour. It implies that there is a consistent set of behaviours for a given system's goal. This is unlikely; human beings are adaptive in their behaviour and need to be because controlled systems are complex and unpredictable. The reference to 'temporal proximity' is also problematic, because it gives no indication of what this means—a minute, an hour, a day? Moreover, the 'immediacy' of the purpose is misleading, because behaviour is fashioned by higher order purposes as well as by more immediate considerations. Generally, such definitions are imprecise, and there is a tendency to use words interchangeably. Thus, in different approaches, 'goal', 'task', 'action' may mean the same (imprecise) thing; behaviour becomes confounded with demands. Stammers *et al.* (1991:138) argue that 'the ambiguity of what constitutes a task is probably one of its greatest strengths, as it provides a flexible framework within which the analyst can group user/operator actions to fit the task context and the overall objectives of the analysis'. While this ambiguity fairly reflects how task analysis is usually practised, it limits development of a more rigorous discipline.

This paper will adopt the perspective of a 'task' as a *problem* facing an operator, with task analysis serving to identify strategies by which the operator can be facilitated to solve the problem, by information design, task re-organization, training and any other human factor. Problems in experimental psychology, such as the 'two-string' problem, the 'selection' task, and the 'Tower of Hanoi' problem, each entail a goal to be met; a representation of an initial state; a set of response facilities with which to affect the state; and an environment or context that constrains the manner in which changes can be made. A specification of a problem is not synonymous with the behaviour that might be used to solve it, but clear problem specification is critical to a proper understanding of behaviour. This paper, therefore, distinguishes between *task* and the *behaviour* recruited to carry out the task.

The aim of this paper is to review the status of HTA and argue its relevance to analysing 'cognitive' tasks. This will be done by providing a justification for HTA as a framework for neutral examination of tasks. Its relationship to analysing cognitive task elements will be justified in terms of identifying occasions when cognitive analysis methods or cognitive modelling might be adopted, and in specifying the demands of the task such that it is clearer how cognition is recruited to meet task goals.

2. Justification of HTA

Annett and Duncan's (1967) ideas of task analysis included elements of task description, behaviour modelling, risk assessment, hypothesis generation and cost-benefit analysis, and were strongly influenced by systems thinking. HTA was justified in part by systems thinking itself, and in part by ideas in skills psychology emerging from systems thinking. A system is any complex of interrelated parts; systems come in many varieties and possess many different properties. A principal characteristic of a system is its function, that is, the purpose the system serves in its broader context. A second important feature is a specification of system components or sub-systems and how they fit together. A system can only be properly understood in terms of its sub-systems by indicating how information, energy resources and physical entities flow between them. Third, expanding a system description by redescribing its components in more detail implies that systems can be described in terms of a hierarchical structure. Generally, the level of detail with which a system should be described depends upon the purpose to which the description will be put. An important property is that if some parts of the system warrant detailed attention, they may be examined in greater detail than other parts of the system.

A hierarchy could be employed simply to describe a system at an appropriate level or it could be used to imply some degree of control. In a hierarchical control system higher levels in the hierarchy exert influence over the manner in which lower levels behave. This may be done in an *authoritarian* manner by passing, to the lower levels, the methods by which the lower level components must be controlled. A less stringent form of hierarchical control is one that is *delegatory*—sub-goals, rather than methods, are assigned to components along with the authority to achieve the sub-goals by whatever method the component deems to be appropriate. Here, the role of the superordinate is to indicate when each subordinate should act. In this form of hierarchical control higher order components maintain their control through monitoring the performance of subordinate components via feedback. At any level, a goal is met by manipulating its sub-functions, while the sub-functions, in turn, work by organizing their sub-functions. An analogy is with structured programming, where sub-routines call on sub-routines and so forth. A further hierarchical arrangement is simply *descriptive*. It implies no control whatsoever, but merely describes sub-goals in relation to one another in terms of their purpose.

Systems ideas were adopted to model human behaviour, especially in disciplines concerned with examining purposeful activity such as industrial psychology or ergonomics. Human beings are regarded as working purposefully but may not share goals with the systems in which they are employed due to a deliberate lack of co-operation, incompetence or because circumstances prevent effective performance. It is the role of human factors engineering and human resource management to ensure that human goals are consistent with the goals of the system in which they are employed. Task analysis is intended to support this objective.

The concepts of feedback and control are strongly represented in modelling human behaviour. Pursuit of a goal entails adapting behaviour to influence a controlled environment in order to attain that goal (Winstein and Schmidt 1989). There are different forms of feedback that can exercise control over human activity, including monitoring limb movement, the feel of controls, the consequences of the action of tools on the environment and comments from other people concerning the outcome of actions. More generalized system feedback is often necessary to help the operator to perform at a strategic or planning level. Miller *et al.* (1960) proposed the

TOTE unit as a model to describe the relationship between action and feedback in controlling performance. The skilled operator: tests (T) the environment against a specified criterion; then, if the test indicates a mismatch, operates (O) to affect change; then tests (T) again to determine whether the discrepancy has now been overcome; then exits (E) when the match is accomplished; if the match has not been accomplished, then (O) is repeated. Annett and Duncan (1967) incorporated the ideas of feedback and control in their approach to task analysis, representing components as input (I), action (A) and feedback (F). Failure to carry out an operation successfully will be due to failure of handling I, A or F.

TOTE and I-A-F can also be compared to 'production systems' used in a number of cognitive modelling approaches (Anderson 1983). A production is a condition-action statement; the condition component corresponds to 'T' (or 'I' and 'F'); the action component of the production rule corresponds to 'O' (or 'A'). A particular production will be triggered when prevailing circumstances match the conditions component—when the condition consistent with a specified pattern is present the action should follow. With a production system, feedback can be dealt with simply by virtue of a different set of prevailing conditions warranting a different action. Clearly feedback is a crucial component of human goal-seeking performance since responses need adjustment in accordance with circumstances in the world. Moreover, it is clear that people can be helped to learn by manipulating feedback during training and helped to perform by providing extrinsic task feedback to help them to regulate their actions. Thus, a production system handles feedback elegantly, but not necessarily as usefully in terms of supporting human factors design. The more explicit statement of feedback in I-A-F and TOTE seems more practical.

A hierarchical skills model that embodies the features of feedback and control is attractive in that a variety of apparently complex behaviours may be represented by relatively simple structures. For a hierarchical model to work, its designer must indicate how components relate to one another to meet the conditions of their common goal. Miller *et al.* (1960) demonstrated that O (operation) could itself be represented as a collection of TOTEs by utilizing a planning component. Annett and Duncan (1967) similarly adopted the word 'plan' as the label for the co-ordinating unit in HTA, although it was only later (Duncan 1972, Shepherd 1976) that plans became necessary components in redescribing goals in HTA and their full benefits became apparent. The utility of the hierarchical description is further in evidence when one addresses the question of how much detail is sufficient in the description of behaviour. Annett and Duncan (1967) proposed the $P \times C$ rule as a means of focusing attention where it is warranted. If attainment of a particular sub-goal is deemed satisfactory in terms of the product of the probability of inadequate performance (P) and cost of inadequate performance (C), then nothing more need be done in terms of task analysis to help people to achieve that sub-goal. Thus, detail in HTA is dictated less by considerations of behaviour than by consideration of *criticality* in terms of the system under scrutiny.

HTA may be seen to represent different things. It may be regarded as a model of human behaviour in the manner just described; this would suggest that the control of observed behaviour corresponds to the hierarchical control implied by different plans in the hierarchy. Supporting this perspective is extremely problematical. It suggests the way in which sub-skills are organized and encoded and implies that plans may be treated as psychological entities. It implies that utilization of information in the pursuit of specific goals is consistent, irrespective of circumstances, and that everyone organizes task data in the same way. It takes no account of the fact that the mental organization of task information would appear to change as people become more or less skilled.

An alternative view is that HTA relates to the strategy of *examining* a task within a practical project. If a goal is met to a satisfactory standard, then the behaviour of people contributing to that goal is assumed to be satisfactory with no further attention warranted. If system performance is judged to be unsatisfactory then contributing behaviours may warrant investigation. One method of investigating this behaviour is through some form of behavioural analysis, including cognitive modelling. Another method is to specify the goal in greater detail. HTA utilizes the strategy of *redescription* into subordinate *operations* and a *plan* that governs the conditions when each subordinate should be carried out. An alternative to investigating behaviour or redescribing a goal is to propose a known solution to that class of problem.

This perspective of systematically examining tasks to establish hypotheses to enable goals to be accomplished, most fairly represents HTA in practice and is consistent in most respects with the original conception of HTA. Thus, HTA is far from an exclusive method and actually relies upon other methods and perspectives being adopted to enable completion. The need for a systematic strategy for task analysis becomes particularly important in large, multiple activity tasks, with which more focused methods, such as *link analysis*, *time-line analysis* or *cognitive task analysis*, could not begin to cope. These more focused methods need a framework to justify their application and to provide sensible boundaries within which they may realistically be applied.

3. Strategy in task analysis

A full task analysis can be a complex enterprise entailing many decisions and judgements carried out over a long period. Figure 1 sets out a process model of task analysis. A variation of this figure was proposed and discussed in Shepherd (1995). Working through this process results in the hierarchical structure of HTA.

3.1. *The HTA framework*

3.1.1. *Box 1—Setting goals*: Task analysis should focus effort in order to be effective and economical. Task analysis should proceed by identifying the main work goal associated with the problem, whether the analysis is prompted by an operational or human performance problem or by a desire to seek improvements to a system. There are several task analysis methods that assist this focusing, for example, the critical incident technique (Flanagan 1954).

3.1.2. *Box 2—Observing constraints*: As goals are discussed, so constraints associated with their attainment or their solution are encountered. Constraints are particularly important in practical projects as they affect the design options that might be adopted to realize goals. Constraints include detail about the work environment that are assumed to influence performance. They also include limitations on preferred solutions imposed by management and staff.

3.1.3. *Box 3—Calculating criticality*: Task elements need only be examined if they are judged to be critical. *Criticality* is a combination of the importance of a goal and

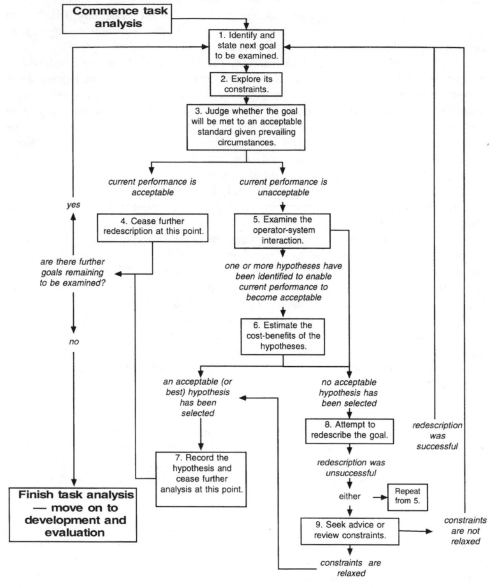

Figure 1.   The Hierarchical Task Analysis framework. This sets out the main stages in a task
analysis and results in the representation of operations and plans that define HTA.

the reliability of its execution, for example the P × C rule (Annett and Duncan 1967).
Analysts may make these judgements subjectively, or they may seek data to quantify
their judgements.

3.1.4. *Box 4—Ceasing redescription*:   An outcome of the criticality judgement in
box 3 is that redescription can cease at this point. This provides one of the important
economies in HTA.

3.1.5. Box 5—Generating hypotheses: If current performance is judged to be unsatisfactory the analyst may examine the operator-system interaction to establish a cause or a solution to the problem. Different project aims may require hypotheses to be stated in terms of *potential causes* or *design suggestions* to overcome the problem. Thus, the discipline of human error analysis tries to identify ways in which human performance might lead to error (Kirwan 1994). *Operational task taxonomies* or *guidelines* may be used to identify design hypotheses without requiring any detailed modelling of behaviour. In other projects where suitable design hypotheses are not obviously forthcoming, the analyst might try to model behaviour as a precursor to suggesting a solution to overcome performance weaknesses. Cognitive task analysis methods may be used to serve this end. In practice, a variety of such strategies may be entirely appropriate.

An important aspect of generating hypotheses is observing constraints. Constraints are identified in box 2 as each cycle of reviewing goals and sub-goals is undertaken. It is quite possible that no hypotheses consistent with these constraints is forthcoming. This will warrant redescription (§3.1.8) or challenging the constraints (§3.1.9).

3.1.6. Box 6—Cost-benefit analysis: If the analyst generates a design hypothesis, then there is a need to determine whether its potential benefits outweigh its costs. This may be done intuitively or may be done explicitly. Some form of *cost-benefit analysis*, suited to routine application within task analysis, needs to be applied quickly, efficiently and repeatedly. Of particular concern is that cost-benefits change as an analysis progresses and a greater number of opportunities for deferring costs present themselves. For example, justifying a training simulator in terms of the first relevant sub-goal to present itself, may be difficult, but as further opportunities to use the device emerge, so cost justification is demonstrated.

3.1.7. Box 7—Recording the analysis: Record keeping shows what progress has been made, how the task has been represented and what design hypotheses have been proposed. It may also be useful to record the hypotheses that have been rejected on cost grounds, since these decisions may be revised as new issues emerge. HTA is recorded through both diagrams and tables (Shepherd 1995). Diagrams are useful in showing task structures but do not show the greater detail of information collected and insights gained.

3.1.8. Box 8—Redescription: Design hypotheses, consistent with cost-benefit criteria, may fail to emerge from examination of the operator-system interaction. Moreover, the task may be judged to be too complex for fruitful application of a formal method of modelling. In these cases the analyst would try to examine the system in terms of its sub-systems. In HTA, exploring sub-systems is achieved through *redescription* by stating a set of subordinate operations and a 'plan' specifying the conditions when each is appropriate.

3.1.9. Box 9—Challenging the constraints: If no hypotheses can be established, the constraints may need to be challenged. At the outset of the task analysis project, management might have ruled out investing in new technology on cost grounds, thereby limiting the options that may be pursued in the task analysis. However, a suitable design hypothesis may not be forthcoming. The following example refers to a case where management sought improvements to the skills of controlling an ageing batch process plant in order to improve productivity. A common requirement in process control is that a particular parameter is required to be maintained at a particular value by adjusting an associated parameter. For example, the formulation of a liquid in a vessel may drift off target over time and must be adjusted by the appropriate addition of a particular feed-stock. In older generations of plant there are still many instances where human operators are expected to carry out this sort of adjustment manually. The task analysis carried out within the constraint of current technology had to focus on training solutions. The skills identified included taking samples, calculating the adjustments to formulation, making the adjustment, then waiting to re-sample and further adjust if necessary. Despite the preferences of management, analysis of these task elements showed that there were no suitable training solutions to meet the standards required, because plant dynamics were not predictable and additional work demands required staff to direct their attention elsewhere during crucial stages. Therefore, the previously stated constraint on changing technology had to be challenged if any productivity improvements were to be forthcoming. To do this, the analyst moved up the hierarchy to a suitable level of description, challenged constraints, and sought different methods of operation that involved automation of control loops.

3.2. *Analytical strategy and the hierarchical representation of the task*
The result of following the processes of review and redescription set out in figure 1 is a hierarchy of operations and plans, recognized as HTA (as in figure 2). Essentially, this is a record of a *process*. Thus HTA does not need to be justified as a model of behaviour, even though there will be occasions when the analyst chooses to use this representation as a model of behaviour, in information requirements specification, part-task training design or the design of documentation, for example.

The above discussion has distinguished between *task* and the *behaviour* that will enable the task to be carried out. For a successful human-task interaction, task and behaviour will complement one another. Tasks or task affordances may be adjusted to suit the capabilities of behaviour; behaviours may be modified to meet the demands of tasks. Task analysis, it is suggested, is a process of examining the human-task interaction in order to establish methods to ensure an appropriate 'fit' to meet system requirements, or to identify reasons why there is not an appropriate fit in a given operational configuration.

4.   Analysing cognition within the task analysis context
Cognition is a crucial aspect of behaviour in tasks especially in connection with those associated with supervision and control of complex automated systems and with human-computer interaction. That specialist cognitive task analysis (CTA) methods should be developed and used throughout the course of a practical task analysis project is becoming an orthodoxy, but it is one that should be challenged on the grounds of economy of effort and the need properly to understand cognitive task elements in the wider task context.

Preece *et al.* (1994:417) have drawn a distinction between CTA and HTA, representing HTA as 'concerned with establishing an accurate description of the steps that are required in order to complete a task' whereas the focus in CTA is to 'capture some representation of the knowledge that people have, or need to have in order to complete the task'. They point out that some actions are physical, such as

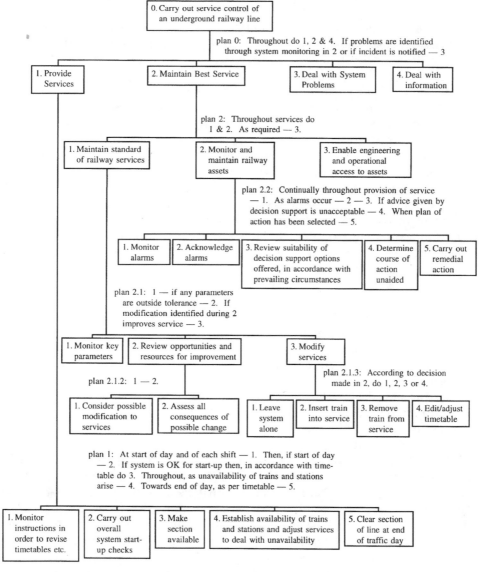

Figure 2. Part of the Hierarchical Task Analysis of supervising an underground railway system. Each of the operations recorded here is further redescribed.

pressing buttons, and some are mental or cognitive, such as deciding which button to press or when to press it. Despite their distinction, successful performance in *all* tasks depends upon the interaction between physical and cognitive elements. Moreover, in a practical project, if part of a task is not critical, then it does not warrant attention, no matter how interesting it is from a cognitive perspective. Rather than distinguishing between cognitive and non-cognitive task analysis one might more profitably consider how a general task analysis strategy accommodates tasks where performance is driven substantially by cognitive factors.

Several earlier attempts at producing CTA methods use task decomposition, similar in structure to HTA, but entailing assertions about how parts of the task description are represented. Thus a GOMS analysis (Goals, Operations, Methods and Selection rules; Card *et al.* 1983) requires the analyst to identify the rules for *selecting methods for* organizing *operators* to achieve *goals*. TAG (Task Action Grammar; Payne and Green 1986) entails coding actions according to rules of syntax to facilitate making comparisons between human-computer interfaces, for example. TAKD (Task Analysis for Knowledge Description; Diaper 1989) entails utilizing rules for knowledge elicitation against task descriptions. In several respects there are similarities between HTA and each of these other approaches. HTA deliberately avoided adopting constraining notations in order to maximize flexibility and to ensure that task descriptions could be understood equally by operational personnel and human factors staff. However, such notations could be incorporated in HTA should the analyst choose to do so. Thus, in different task analysis projects, the analyst might recognize the benefits of standardizing levels of description (as in GOMS) or expressing operations according to a particular notation (as in TAG) or recognizing that an important implication of plans is that there is a declarative knowledge base that may be generated to facilitate people learning a new task (as in TAKD). This is not to claim that HTA does what these other methods seek to do, but merely to point out that common forms of description are employed in different methods and that HTA may also be adapted in these ways. The important point about CTA methods is that their authors are attempting to exploit principles of cognition in their choice of constraints on task descriptions regarding, for example, the means by which syntax is parsed or the manner in which declarative knowledge is compiled to acquire skill. These methods attempt to do different things from one another and should be used by the analyst for different purposes, as the task analysis dictates. Important benefits of incorporating CTA within the HTA framework are economy of effort and in ensuring that the examination of task elements is not carried out in a vacuum.

More recently, approaches to cognitive modelling in support of CTA have attempted to develop what Kjaer-Hansen (1995:43) describes as 'unitary theories of cognitive architectures'. These are more global theories that aim to understand cognition in terms of the role of knowledge within the processes of learning, inference and decision-making in producing behaviour, for example, SOAR (Laird *et al.* 1987) and ACT* (Anderson 1983). From a performance perspective, the unitary theories would account for the manner in which information is acquired, stored, organized and retrieved to account for decision-making and performance across a range of task elements. This perspective is important because despite the fact that tasks can be described in terms of separate goals, cognition is more widely influential and HTA frequently demonstrates the need to account for behaviours that interact across the task hierarchy.

One may distinguish between *cognitive task elements*, representing the problems with which operators or users are required to cope, and *cognitive skills*, concerned with operating upon information gained from the environment in order to guide subsequent physical actions. There may be no simple mapping of cognitive task elements onto cognitive skills, yet it is essential that task elements must be identified before skills can be examined. Otherwise it is unclear what these skills must achieve and there is no guidance for evaluating whether or not skills have been achieved.

Physical task elements are usually observable, whereas cognitive elements must be inferred by their relation to the observable actions of the task. Systematic variation of subsequent physical action implies that a decision has been made. The analyst needs to reflect carefully on how task elements interact before modelling underlying cognitive skills. In a nursing context, for example, engagement with the patient during routine care may provide crucial information for dealing with a subsequent crisis. Therefore, decision-making and routine care tasks cannot sensibly be analysed without reference to each other.

4.1. *Inferring cognitive task elements*
Cognitive task elements can be inferred through HTA by stating plans. If different actions are consistently carried out in different circumstances, then a *decision* process can be inferred that can feature as an operation as part of a redescription, even though decision-making was not apparent through observation or through talking to the operator. Similarly, if following a period of apparent inactivity an operator is seen to respond swiftly and appropriately to an unforeseen set of circumstances, it follows that the operator was *monitoring* a signal. A common work requirement is to maintain the status of a system—a refinery operator may 'maintain correct product specification'; a railway supervisor may 'maintain train headways'. To carry out these tasks implies that the target is monitored and then action is taken to rectify any deficiency. Remedial action could entail referring the problem to a colleague or making a simple correction. Alternatively it could entail a combination of *diagnosis*, *rectification*, *compensation* and *recovery*.

By inferring cognitive task elements within the context of a general task analysis it is also possible to identify what it is supposed to achieve. A monitoring task or decision-making task may vary considerably in accordance with demands placed on the person carrying out the task. This may be illustrated by reference to the case of a railway systems engineer, employed to monitor the health of the railway system in terms of trains running to timetable and whether assets (trains, track, escalators, etc.) are functioning correctly. If a problem arises the engineer needs to decide on a course of follow-up action. The nature of this decision depends on the courses of action permitted by management. The task is undertaken in a central control room, with maintenance teams deployed to undertake the work at the affected site. The demands on monitoring and decision-making will vary in accordance with the organization of work. In one form of work organization versatile technicians may deal with all the technologies of the railway—mechanical, electrical, electronic. The systems engineer would undertake an initial diagnosis, then brief a maintenance team. The initial diagnosis cannot be complete, because the central control room only enables limited access to information; inspection of local conditions is also necessary and this must be done by the maintenance team when they reach the site. Therefore, the engineer's task is one of reduction of uncertainty to facilitate the search that the team will need to carry out when it reaches the site. In a second form of work organization maintenance teams are not multi-skilled but specialize in particular technologies. Here, the engineer must deploy the team most likely to deal appropriately with the problem. Deploying any team is costly. Should it transpire that the initial judgement was wrong, then these costs will have been wasted and the next hypothesis selected. Moreover, the judgement of whom to deploy will be a function of availability, costs, likely search time, likeliness of solutions and the severity of the incident. These judgements will be far more demanding than the simpler hypothesis reduction needed in the first form of work organization where teams were multi-skilled. A further variant is that *contractors* are used to deal with all maintenance; the systems engineer merely monitors the contractors' response to events. This entails noting incidents, then awaiting telephone calls from the contractors' management to say that their teams are responding. The engineer's job would then simply be one of quality monitoring to ensure that the terms of the contractors' contract are being observed. Each of these variants entails monitoring and decision-making, but associated cognitive skills will be trivial or complex in accordance with the decisions and actions that follow. The task must be first understood to anticipate the demands placed on cognition.

Just as consideration of actions generates cognitive elements, so identification of cognitive elements points to actions to be included in the task analysis. An operator may report that, after '*checking* the melting point of a substance', a standard fixed procedure is followed; observation of several instances of the task may appear to confirm this fixed procedure. However, *checking* or any other information gaining activity implies at least two outcomes—the system status is satisfactory or it is unsatisfactory. The invariant routine observed may simply reflect a reliable system where the unacceptable outcome is rarely encountered. If, on the other hand, remedial action is *never* necessary, then the checking action can be ignored. The only occasions when checking a parameter is a legitimate part of a *fixed* procedure is when the outcome is recorded for use by somebody else, for example, quality control. Task analysis should establish the interaction between cognition and action before making assumptions about the nature of cognition or, indeed, the necessity for action.

4.2. *Inferring cognitive skills*

Identifying a cognitive element in task analysis implies an underlying *cognitive skill*. Sometimes this skill is only concerned with a single task element identified through the task analysis. However, sometimes a more general skill must be understood to deal with the interaction between several task elements. In the tasks associated with delivering neonatal intensive care, represented in figure 3, each of the 11 sub-operations entails cognition of some kind—judgement, planning, or problem solving. Operation 1 entails *procedures*, but these may vary according to the condition of the baby. Operation 2 entails *diagnosis* and *planning*. The diagnosis component may entail collecting information to enable the diagnosis and selection of treatment. As treatment is delivered it is possible that the current treatment/care plan ceases to be suitable and needs revision (operation 4). This implies that a judgement is made continually to assess the adequacy of the current plan, hence operation 3 is inferred. The remaining operations may be treated in a similar manner. If the cognitive skills associated with each task element were examined separately it would not reflect the proper nature of the task. If a carer is responsible for all of these task elements in the care of a baby, then information may be gained deliberately or incidentally at various stages, and possibly stored for later use. Information is threaded throughout all aspects of this task to support planning and problem-solving. The skills and expertise of care, or indeed any systems supervision and control task, would need to be understood in terms of a more general cognitive system involving memory, inference, problem-solving and planning. It is for the conscientious task analyst to recognize these inter-relationships between the skills underlying task elements to avoid an oversimplification of the problem.

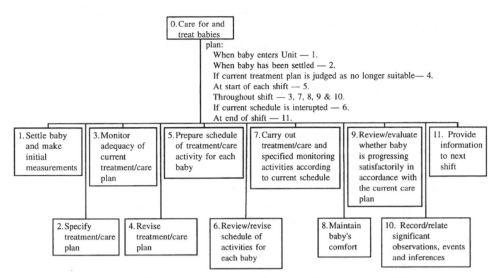

Figure 3. The top level description of the task of carrying out neonatal intensive care.

4.3. *Cognitive task elements in context*

Discussion of the railway system engineer's task in § 4.1 indicated the importance of considering subsequent action in characterizing a decision-making task. There are further factors that influence performance of a cognitive task element.

4.3.1. *Goal context*: Goal context refers to the wider goals within which a task element of interest resides. Figure 2 represents the task of supervising the operation of a railway system. This is a complex task in the sense that there are many different things to do, several of which are interrelated. The individual task elements are not necessarily complex in themselves. The job is to ensure that the railway is running safely: timetables and reasonable distances between trains are maintained; problems are identified and dealt with; information is communicated to interested parties around the railway network. Taking one example, 'Assess all consequences of possible changes {2.1.2.2}' is carried out to contribute to 'Review opportunities and resources for improvement {2.1.2}', which is part of 'Maintain standard of railway services {2.1}', which is part of 'Maintain best service {2}', which is part of 'Carry out service control on the line {0}'. The performance of two people carrying out this task will differ if their wider goal contexts differ. Thus, a supervisor standing in to help overloaded staff will be operating in a different context to a signalling assistant. The supervisor may adopt a strategy that supports immediate systems goals as quickly as possible to resolve the immediate congestion; the signalling assistant, who will still be on hand to deal with the consequences of the action when the crisis is over, may choose to act in a way to minimize work during a subsequent recovery activity.

4.3.2. *Frequency, predictability and coincidence*: The task analysis should indicate where frequency and predictability of events affect performance. The plans in HTA specify the cues for action; the actual occurrence of events will indicate the frequency and predictability of these cues. This influences the choice of human factors solution. Response to predictable events can be automated, or proceduralized—at least personnel will be well prepared for these events and be on hand to respond. An uncertain task brings stress from uncertainty and, combined with infrequency, limits opportunity for practice. Such tasks are also often critical, for example, diagnostic tasks. Taking a task out of context of time and expectation may destroy much of what is important about the task.

Where events are unpredictable, several might occur together, causing extra workload. The HTA structure indicates, through the plans, where these coincidences might occur. For example, in figure 2, 'Make section available {1.3}' could be required at the same time as 'Establish availability of trains.⋯{1.4}', 'Modify services {2.1.3}', 'Review suitability of decision support options⋯{2.2.3}', and many other combinations, because they are all driven by events that might coincide. Unless job-design solutions include increasing the size of operating teams, the underlying cognition required to cope with these events in combination may be insufficiently robust. Plans in HTA, together with a record of events, will indicate the performance criteria to be met.

4.3.3. *Priming*: If the operator is oriented towards a set of circumstances by just having carried out activities that always herald the new task demand, then performance will be *primed*. Some priming is *procedural* in the sense that it prompts the operator what to do next, and is made explicit in plans. Some priming is *informational* in that attainment of previous goals may provide information to the operator, which will be of use in later activities. Thus a doctor or nurse engaged directly in administering care or treatment would have a more thorough working knowledge of the patient than a staff member not similarly engaged and this is likely to affect diagnostic strategy in dealing with any crisis that might subsequently occur. Examining diagnostic skill out of context may be inappropriate.

4.3.4. *Decision outcome*: The issue of decision outcome has already been discussed in § 4.1 in relation to the railway systems engineer's task. A diagnostic problem where the operator must distinguish between which of several specialist maintenance technicians need to be alerted is very different to one where the operator is required to determine the course of remedial action in terms of which pieces of equipment need replacing, and is very different to a binary choice of keeping a system operating or shutting down.

4.3.5. *Criticality*: Task criticality, discussed in box 3 in § 3.1.3, relates to the criteria of performance to be observed in any study of cognition and to whether attention to this particular task element is, indeed, justified.

4.4. *Evaluating cognitive skill*
A final necessary link between CTA and a form of task analysis that records operational requirements, such as HTA, is that practical insights gained from CTA, either problem hypotheses or design hypotheses need to be evaluated. Such evaluation must be informed by task requirements in terms of work goals and system events. Specification for such an evaluation will be derived from a task analysis that records system demands, rather than simply through modelling cognition.

5. Conclusions

The paper has considered two interrelated themes: (1) the justification of Hierarchical Task Analysis as a framework for analysing tasks, rather than as a method for modelling behaviour; and (2) the treatment of cognitive task elements within this framework.

The word 'task' is defined in terms of stating the problem that people with system responsibilities are charged with resolving. Thus task is treated as complementary to behaviour and not synonymous with it. Task analysis, it is suggested, is a process of examining the human-task interaction in order to establish methods to ensure an appropriate 'fit' to meet system requirements, or to identify reasons why there is not an appropriate fit in a given operational configuration.

Many developments in task analysis have added to the plethora of methods, requiring analysts to decide which methods to apply on different occasions. It is argued that making such choices should be integral to task analysis and not a precursor to it. A task analysis framework is presented that relates the various stages of decision-making encountered in a practical project, including identifying the task to be analysed, exploring constraints on solutions, exploring behaviours, cost-benefit analysis, and task redescription. Working through this framework results in the hierarchy of operations and plans that characterizes Hierarchical Task Analysis. Thus, Hierarchical Task Analysis is justified as a general task analysis strategy rather than as a method for modelling behaviour.

All performance entails an interaction between cognition and action, so it is difficult to understand why some tasks are assumed, from the outset, to warrant a specialist form of cognitive task analysis, while others are not. Cognition should be treated within a general task framework, such that it can be understood in relation to action. Should a specialist method of cognitive task analysis or cognitive modelling be required in examining any particular aspect of task behaviour, then this will become clear as the task analysis develops. Identifying cognition within the context of a neutral task analysis will situate it within a task goal, it will indicate the extent to which task elements are carried out alongside other demands, it will indicate which activities influence performance by having gone before and which choice of actions will follow. Moreover, considering cognition within the context of a real task will indicate whether effort to examine it in detail really is justified.

Acknowledgements

Task analysis associated with Neonatal Intensive Care was carried out as part of EPSRC Grant GR/K88590. Grateful acknowledgement is also due to staff at various Neonatal Intensive Care Units and staff of London Underground in providing information to enable task analysis to be carried out.

References

ANDERSON, J. R. 1983, *The Architecture of Cognition* (Cambridge, MA: Harvard University Press).

ANNETT, J. and DUNCAN, K. D. 1967, Task analysis and training design, *Occupational Psychology*, **41**, 211–221.

ANNETT, J., DUNCAN, K. D., STAMMERS, R. B. and GRAY, M. J. 1971, *Task Analysis* (London: HMSO).

CARD, S. K., MORAN, T. P. and NEWELL, A. 1983, *The Psychology of Human-computer Interaction* (Hillsdale, NJ: Lawrence Erlbaum).

DIAPER, D. 1989, Task analysis for knowledge descriptions (TAKD); the method and an example, in D. Diaper (ed.), *Task Analysis for Human-Computer Interaction* (Chichester: Ellis Horwood).

DUNCAN, K. D. 1972, Strategies for the analysis of the task, in J. Hartley (ed.), *Programmed Instruction: An Education Technology* (London: Butterworth).

DUNCAN, D. K. 1974, Analytical techniques in training design, in E. Edwards and F. P. Leeds (eds), *The Human Operator and Process Control* (London: Taylor & Francis), 283 – 320.

FLANAGAN, J. C. 1954, The critical incident technique, *Psychological Bulletin*, **51**, 327 – 358.

KIRWAN, B. 1994, *A Guide to Practical Human Reliability Assessment* (London: Taylor & Francis).

KIRWAN, B. and AINSWORTH, L. K. 1992, *A Guide to Task Analysis* (London: Taylor & Francis).

KJAER-HANSEN, J. 1995, Unitary theories of cognitive architectures, in J.-M. HOV, P. C. Cacciabue and E. Hollnagel (eds), *Expertise and Technology: Cognition and Human-Computer Cooperation* (Hillsdale, NJ: Lawrence Erlbaum).

LAIRD, J. E., NEWELL, A. and ROSENBLOOM, P. 1987, SOAR: An architecture for general intelligence, *Artificial Intelligence*, **33**, 1 – 64.

MILLER, G. A., GALLANTER, E. and PRIBRAM, K. 1960, *Plans and the Structure of Behaviour* (New York: Holt, Reinhart & Winston).

MILLER, R. B. 1953, A method for man-machine task analysis. *Report WADC-TR-53-137*, Wright Air Development Center, Wright-Patterson AFB, Ohio.

PAYNE, S. J. and GREEN, T. R. G. 1986, Task-action grammars: a model of the mental representation of task languages, *Human-Computer Interaction*, **2**, 93 – 133.

PREECE, J., ROGERS, Y., SHARP, H., BENYON, D., HOLLAND, S. and CAREY, T. 1994, *Human-Computer Interaction* (Wokingham: Addison Wesley).

SHEPHERD, A. 1976, An improved tabular format for task analysis, *Journal of Occupational Psychology*, **49**, 93 – 104.

SHEPHERD, A. 1995, Task analysis in HCI tasks, in A. F. Monk and N. Gilbert (eds), *Perspectives in HCI* (London: Academic Press).

STAMMERS, R. B., CAREY, M. S. and ASTLEY, J. A. 1991, Task analysis, in J. R. Wilson and E. N. Corlett (eds), *Evaluation of Human Work* (London: Taylor & Francis).

WINSTEIN, C. J. and SCHMIDT, R. A. 1989, Sensorimotor feedback, in D. H. Holding (ed.), *Human Skills* (Chichester: Wiley).

CHAPTER 3

Integrating task and data flow analyses using the pentanalysis technique

D. DIAPER, S. McKEARNEY and J. HURNE

Department of Computing, School of Design, Engineering and Computing,
Bournemouth University, Poole House, Talbot Campus, Fern Barrow, Poole,
Dorset BH12 5BB, UK

Keywords: Task analysis; Data flow analysis; Human-computer interaction;
Software engineering; Pentanalysis technique; Data flow diagrams; Task analysis
for knowledge descriptions.

Pentanalysis is a new method that allows the integration of task analysis with data
flow analysis. This is novel and desirable as it provides one means by which a
human-computer interaction orientated method can provide input to the
requirements specification stage of the software life cycle. It bridges the gulf
that exists between this stage and one major representation, data flow diagrams,
commonly used in the software design stage. The method is described and a
worked example is provided to illustrate it.

1. Introduction

In a well ordered universe software engineering and Human-Computer Interaction
(HCI) would not be separate disciplines. They share a common goal of improving
how computers are made and used and a common, overall perspective of being
engineering disciplines, rather than being either science or craft (Long 1986, Long
and Dowell 1989). Both disciplines are relatively new and share a similar historical
imperative; software engineering developed in the late 1960s in response to the
'software crisis', where customer demand for software was recognized to exceed
production capability, and HCI emerged in the late 1970s in response to the demand
for computers to meet user needs.

The legacy of these two disciplines' development remains evident today in the
difference in culture between the two. Software engineering takes a computer
perspective of situations whereas HCI is inherently anthropocentric. This is not to
suggest that a great deal of software engineering is not about people; it is, for
example, in its concern for project and personnel management, or that HCI does not
have a major computer-related element. Overall, however, software engineering and
its practitioners tend to have a computer-related focus and HCI a psychological or
ergonomic one. Such differences are reflected in how the two disciplines identify
problems and how they attempt to solve them. Diaper (1990a, 1997, Diaper and
Addison 1992a) has previously argued that such a culture difference is particularly
notable in the gulf that he has identified between the specification of requirements
and the specification of design. The former is inherently anthropocentric in that
requirements should specify what people need. Design on the other hand, is
computer centric, that is, it specifies computer system design. Of course, these
distinctions are not absolute in that design must consider the environment of an
intended computer system and this environment certainly contains people, not least,

as end-users. Similarly, requirements specification must consider what is feasible and one aspect of feasibility is a technical, computer-related component.

One of the main classes of solution proposed by the software engineering community has been the development of structured methods for software development. These methods are intended to improve both the quality of software (typically its maintainability, reliability, efficiency and usability: Sommerville 1989) and the productivity of those who produce software by placing the implementation stage, the actual writing of computer programs, in a broader context. Boehm (1975) is widely quoted as estimating that implementation represents only about 20% of the cost of software development and that the requirements and design that precedes implementation and the testing that follows account for approximately 40% each of the total software development costs. These costs vary, of course, with the nature of the software product, the development approach used, the tools available, and so forth, but even so, Boehm's cost estimates for implementation range only from 17% (for operating system development) to 28% (for business systems).

The earliest forms of structured methods were the waterfall models where there is a cascade from requirements analysis, to design, then implementation and finally to testing. Pressman (1987, 1994) calls this 'The classical life cycle' and relates it to the engineering of physical products. Such models have been criticized for many reasons, but particularly because they do not describe what software developers actually do. Realistically, iteration between stages has to be allowed and the waterfall aspect is maintained in that if an earlier stage is subsequently changed then all later stages must also be revisited.

The subsequent development of structured methods has concentrated on the specification of particular representations used at more, and more precisely defined, stages. Jackson Structured Design (JSD) was one of the first such structured methods and used what would now be recognized as Entity Life Histories (ELHs) in a six-step process of which only the final step involved implementation (Cameron 1983). Lim and Long's (1994) Method for USability Engineering (MUSE) was originally based on an extension of JSD, called JSD*, and has a more sophisticated structure of stages and relationships between stages than JSD. The Structured System Analysis and Design Method (SSADM; Downs *et al.* 1988, Ashworth and Goodland 1990, Hares 1990, Eva 1994) has perhaps reached the apogee of such approaches to development with Downs *et al.*, for example, listing 11 'techniques', each with its own representational formalism, and 35 main 'steps' in SSADM. There is considerable opportunity for iteration between many of these steps.

Structured development methods have excited the HCI community for more than a decade because they appear to provide a means by which ergonomic and human factors concerns can be introduced into the early stages of the software development process, in particular, within the requirements analysis stage. While there are notable exceptions, for example Lim and Long (1994), the HCI community have generally assumed that if they can incorporate better human factors into requirements then this must inevitably lead to the production of better computer systems. Diaper's (1990a, Diaper and Addison 1992a) gulf between requirements and design questions this assumption on the basis that better, more user-centred, requirement specifications will only lead to better computer systems if such requirements are actually incorporated into the design stage. His conclusion is that the cultural differences between the software engineering and HCI communities prevents computer system designers from making as much use as they might of the HCI contributions to requirements in real commercial projects.

The purpose of the work reported in this paper is to continue a programme of research that attempts to bridge the gulf between task analysis and traditional software engineering approaches. The overall approach has been to choose the representational formats associated with powerful methods within requirements and design and to provide mechanisms that allow the methods to support each other by providing some form of translation between the representations. Diaper's work has concentrated on using task analysis as a requirements analysis method. This paper reports the success of achieving the goal of relating a task analysis to one of the most widely used design representations in software engineering, Data Flow Diagrams (DFDs). The scope of the paper is modest, however, in that it explicitly does not attempt to change how system designers use DFDs within particular approaches, structured or otherwise, but it does demonstrate how DFDs can be related to the output from an anthropocentric task analysis. Diaper (1997) argues that it is undesirable to attempt to change software engineering approaches with which its practitioners are familiar and comfortable on the grounds that it is software engineers who design and build computer systems and it is more plausible, and likely to be successful, if the HCI community tries to meet the builders needs, rather than trying to change them.

1.1. *Task analysis*

While it is not claimed that task analysis is sufficient, alone, for the specification of requirements, it does provide an extremely powerful and potentially rigorous means of specifying many of the technical requirements of a system based upon an anthropocentric perspective, albeit one that takes the tasks of a human-computer system as its central view. There are many advantages to the task-centred view of the world and task analysis has the capability of representing a wide variety of date types from many sources (Johnson *et al*. 1984a,b).

In general, tasks are the means by which the Interactive Work System (IWS) which includes both people and tools, such as computers, effect changes on the real world, otherwise called the application domain (Long 1989). Thus the IWS model of tasks is that it is not people or computers alone that carry out tasks but their collaboration that achieves changes in the world.

An early attempt to use task analysis to provide design-centred representations was the method of Task-Action Grammar (TAG), which was designed to generate Bachus-Naur Form (BNF) as its output (Payne and Green 1986, 1989). Unfortunately, BNF is no longer widely used as a representational form in software engineering and TAG has only ever been demonstrated as suitable for the analysis of low level user interface design features.

The Task Analysis for Knowledge Descriptions (TAKD) method was originally developed to specify training needs by identifying the common components of a very wide range of tasks and representing these as specifications of the knowledge required for successful task performance (Johnson *et al*. 1984a,b, Diaper and Johnson 1989). TAKD has since been developed as a general method for the capture and analysis of requirements in HCI (Johnson 1985, Johnson *et al*. 1985, Diaper 1989a, 1990b, Diaper and Addison 1991) and for the specification of design options in computer centric representations. For example, TAKD can generate algebraic formal specifications using OBJ (Gikas *et al*. 1990) and, with greater difficulty, formal specifications as Z Schema. More recently (Diaper 1997), TAKD has been shown to be capable of supporting entity modelling in the context of software engineering's structured design methods such as Jackson Structured Programming (JSP) and Jackson System Design (JSD) (Cameron 1983) and SSADM and could equally well be used in HCI rooted structured methods such as MUSE.

The original intention of the research reported in this paper was to further develop TAKD, and particularly its Task Descriptive Hierarchy (TDH) representation to support data flow analysis. This approach was unsuccessful because while a TDH has the desirable property that, like Data Flow Diagrams (DFDs), it does not represent sequence, its hierarchical nature could not be made to fit the non-hierarchical nature of a DFD at a particular level. The authors therefore see the new 'Pentanalysis technique' described in this paper as supplementary to TAKD and in the context of, for example, SSADM, it is believed that both methods can be applied to the same task data to separately support entity modelling and data flow analysis.

Activity lists or task protocols are a common representational form in virtually all task analysis methods (Diaper 1989) and are used as the starting point of the pentanalysis technique. An activity list is merely the prose description of one or more tasks to be analysed. Whatever the initial language style of an activity, it is a common heuristic to produce such a description consisting of a list of short sentences that each contain one action. In addition, virtually all task analysis methods contain some notion of dividing the world into objects and actions (Diaper 1989). While the specification of objects is frequently complex because objects are generally compound in nature and because the number of objects tends to exceed the number of actions by an order of magnitude (Diaper and Johnson 1989), the specification of actions is generally fairly easy. Many analysts merely think of them as the main verb of a sentence describing some task step. Actions, of course, are compound like objects but their relative infrequency is caused because nearly all tasks consist of a small number of actions that are applied to a variety of different objects. For example, actions such as 'type' can apply to anything typed on a keyboard and the meaning of the action is determined by its associated object or objects, e.g. type command, type return, type text, type menu option, etc.

Task analysis, perhaps because of its history in scientific management (Taylor 1911) and its early use in training (Annett and Duncan 1967, Annett *et al.* 1971), has often concentrated only on what people do in tasks. This was, for example, an explicit decision for the design of TAKD in the early 1980s. The authors would now argue that this is taking the user-centred view to a ridiculous extreme, it is so user-centred that it ignores everything in the world except the user. In contrast, the pentanalysis technique is designed to support the design of computer systems by facilitating computer-orientated data flow analyses.

Diaper's (1997) related work to that described in this paper used Long's IWS task model; this caused him to modify the style of the activity list he generated compared to those user-centred ones he had previously generated for TAKD analyses. This led to many lines being mirrored, one for each perspective. For example:

```
user types customer number on keyboard
computer accepts keyboard input customer number
```

These mirrored descriptions of user and computer are based on a co-operative philosophy and an overall view to functional allocation that people should do what they are good at and likewise for computers (Diaper 1989c). The 'one action per line' heuristic naturally forces this twin line structure on the activity list. It is also natural to consider the computer as another agent and in effect treat it in a way similar to how a person is treated in the task analysis. This is consistent with Bench-Capon and McEnry's (1989a,b) view that computers are merely a channel for mediated discourse between end-users and programmers.

1.2. *Data flow analysis*

Data flow analysis is undoubtedly still one of the most commonly used computer-centred analysis techniques in software engineering. The main representational tool of data flow analysis is the Data Flow Diagram (DFD). As well as being one of the three major views of system data in SSADM, DFDs are frequently used in less organized approaches to system design. Superficially, data flow analysis looks simple in that it provides a high level view of how data is processed by the system under study and how the data and processes within the system relate to objects outside the system boundary. One of the many claims made by proponents of DFDs is that they capture the most important features of an information system, i.e. the data flows between processes (Batini *et al.* 1992). Many authors (Cutts 1987, Batini *et al.* 1992) claim that the simplicity of the DFDs makes them ideal for discussing features of the system with the user or client. While the appropriateness of DFDs for systems analysis may be debated, their use in so many software engineering methodologies, including many documented projects, cannot be disputed.

A DFD represents data flows between three types of 'elements' in the system: (1) transformations of data, called processes, (2) static stores of data, called data stores, and (3) data consumers or producers external to the system, called terminators (external entities). As Sommerville (1996:101) says 'There are various notations used for data-flow diagrams' and a simplified notation has been used for illustration purposes, loosely based on that used in structured methods such as Yourdon (1989) and Davis (1990) for figures 1 and 2, which show a simple DFD at Level 0 and Level 1, respectively. Given the many alternative notations available for drawing DFD's depends on the chosen methodology and CASE tool, it was decided not to use the DFD notation of SSADM as the notation contains identifiers for each symbol which, while adding value to a project's documentation, can make the diagram appear unnecessarily cluttered to an unfamiliar reader.

Traditionally, there are a number of constraints on how elements may be connected in a DFD, although different analysis methods apply these constraints differently. Data flows always represent actual flows of information between elements but only some methods represent associated control information. In all data flow analysis methods, data may not flow directly between data stores and terminators but must pass through a process that may generate, destroy or transform data. Data flows into a process, may originate in another process, a data store or a terminator and the data flows out of a process may go to another process, be stored in a data store or passed to a terminator.

Perhaps the most important property of a DFD is that time and, therefore, contingency is not represented in the diagram, i.e. a data flow merely represents that in some circumstances data, of a particular type, is passed in one direction between two elements; when this might occur and under what preconditions is not represented. While it may seem odd to claim that such a lack of constraint is advantageous, it is, because it simplifies the already considerable problem the software engineer has of understanding the real world by ignoring one very complex aspect of it. A similar argument is used in TAKD where its main analytical stage, TDH construction, ignores task sequence data, which is analysed later within TAKD, if necessary.

Data flow analysis uses a top-down, decompositional mechanism involving the identification and progressive expansion of each process in the system. Analysing or designing a system using DFDs generally starts with a 'context diagram', often called a 'Level 0 DFD', that shows the system as a single process with data flows to and from external entities (figure 1). The Level 0 DFD helps to establish the boundaries of the system under study. Diagrammatically, processes are represented by labelled circles and data flows are represented by labelled arrows. Data flows are always one-way. The analysis proceeds by decomposing the single context diagram process into a 'Level 1 DFD' (DFD 1) that consists of two or more processes and, optionally, data stores. The analysis ensures that inputs to and outputs from an expanded process are represented in lower level processes. Hence, while DFDs abstract processes, data flows must be fully represented at each level in the DFD set. For example, figure 2 shows the level 1 DFD for figure 1. Data stores are represented by open-ended rectangles and generally they only appear in a particular level when accessed by more than one process. If necessary, each process in the level 1 DFD can be further decomposed to produce a set of level 2 DFDs. Therefore, data flow analysis produces a hierarchy of individual DFDs.

While data flow analysis is widely described in the software engineering literature, few authors explore the decision-making processes behind the decom-position of DFDs. DeMarco (1979) suggests that all the processes at a single level should be approximately 'even-size' and he uses simple case studies to demonstrate what is meant by this term. Batini *et al.* (1992:114) suggests that sub-processes should be 'well-distinguished' and independent of each other and suggests that a DFD is complete when 'it represents all the features of the application domain at an appropriate level of detail'.

Specific guidelines or rules for decomposing processes are not provided. Batini *et al.* (1992) for example, discuss a number of general techniques for arriving at an appropriate DFD. Diaper (1997) has argued that this lack is hardly surprising as procedures, such as DFD decomposition, are inherently difficult to describe in a declarative form such as natural language. He argues that this is one major reason why the literature generally substitutes worked examples for any serious attempt to describe such procedures in a detailed, context-free fashion.

The software engineer is really relying on heuristics and craft skill to produce a reasonable, usable DFD and it is always possible to produce alternative DFDs as

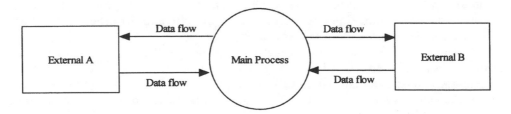

Figure 1.   A level 0 DFD.

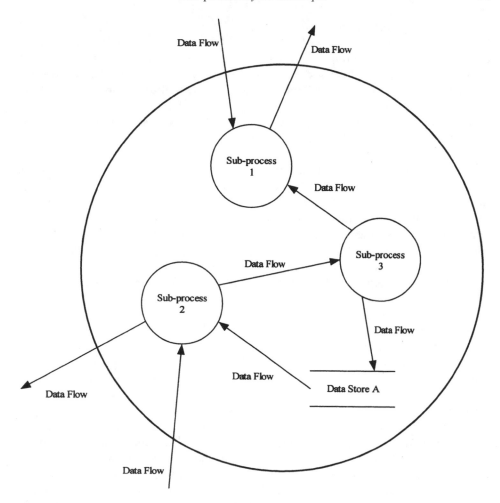

Figure 2. A level 1 DFD.

decomposition proceeds. It can be argued that this is actually a satisfactory state of affairs in that the software engineer is attempting to understand the world and represent some simplified abstraction of it that is useful for the purpose of computer system specification and design. Such tasks are appropriate for people and exploit the power of the human mind as an information processing device capable of dealing with complex, multivariate problems.

While a software engineer generating DFDs is using a deductive logic and ignoring temporal information, in reality she or he tends to consider cases of what might happen in the real world and builds parts of the DFD accordingly. Where this approach is weakest is in that such considerations are usually done *ad hoc*, piecemeal and casually. HCI, and in the context of this paper, task analysis, is in theory capable of supporting the production of DFDs by recognizing this modelling of the world and undertaking it in a more structured, complete and systematic manner.

1.3. *Task and data flow analyses*

Data flow analysis involves the observation and description of a system. The DFD is one method of representing the results of data flow analysis. Most advocates of DFDs accept that it is a heuristic-based process requiring a lot of skill and experience to achieve success. Similarly, task analysis also involves observation and description of a system and, as with DFD development, task analysis methods require a lot of expertise and experience to be carried out successfully. The purpose of this paper is to show how task analysis and data flow analysis can mutually support each other's description of a system.

It is generally accepted within both software engineering and task analysis that there will be considerable iteration between the stages in each of these techniques. In a task analysis method such as TAKD, for example, the primary analysis cycle consists of three stages (activity list description; specific action and specific object selection; and task descriptive hierarchy construction). The first stage, which is common to virtually all task analysis methods (Diaper 1989d), provides a prose description of a task or tasks and as such represents an abstraction of the real world in natural language. That this description changes during the task analysis, usually, but not always, by the addition of further detail, implies that the process of task analysis itself changes the data observed from the real world (Diaper 1989b). In science this would be viewed as cheating since it is a form of *post hoc* data selection. In contrast, in engineering it represents a proper approach to the refinement of a view of the world relevant to an engineering project's goals. A similar process occurs within the iteration between the requirements and design stages as represented in the classic waterfall model of the software life-cycle (Diaper 1989b).

Initially, the task analysis and data flow analysis are carried out in parallel and the consequence of subsequently iterating between the task and data flow analyses, as the redesign of the current system proceeds, is the sort of refinement of these real-world descriptions that typically occurs in most software design processes. As redesign proceeds, but before the new system is implemented, it is possible for the analyst to generate task scenarios of parts of the system that are redesigned and to use these to predict how tasks will be carried out post-implementation. The expectation, of course, is that once implemented, these scenario-based task predictions will be evaluated using the same approach to task analysis used in the requirements analysis.

2. The pentanalysis technique

This section of the paper outlines the main components of the pentanalysis technique. How these components are then deployed, and some of the heuristics that analysts might use, are subsequently described in the worked example (§3).

2.1. *Requirements data capture*

External to the pentanalysis technique is the capture of requirements data. It is assumed that this will involve traditional methods used in both software engineering and HCI and would include some of the following example techniques: task observation, interviews of many types with many people, scenario description, walkthroughs, study of the extant system, goal analysis, specification of system and project constraints, user and other personnel modelling, heuristic analysis, ethnography, surveys, training and operational manuals, etc.

Part of this data capture exercise must precede anything else, but it may occur repeatedly during analysis and it is almost certainly desirable that this is so. Desirable, because it is a potentially expensive process and costs, in terms of people, time and money, will be significantly reduced if such capture is more focused on what is needed for the requirements and design analysis.

2.2. *The pentanalysis technique*

After the initial data capture exercise, the pentanalysis technique then undertakes two analyses in parallel. A traditional data flow analysis and the production of a task analysis activity list. The initial data flow analysis exercise is carried out as normal and produces an initial set of DFDs (e.g. level 0, level 1, etc.). The goal of the method is to integrate both the activity list and DFDs and thus enhance the analysts' understanding of both representations of the system. Figure 3 shows the main stages of the pentanalysis technique.

As discussed earlier, the integration of the activity list and the DFDs will be an iterative process that provides feedback regarding the consistency, accuracy and validity of both representations. While working on the case study the authors frequently found themselves discussing inconsistencies between their perceptions of the data flows in the worked example and the actual activities observed during the task analysis. This resulted in changes in the activity list and the DFDs. This type of relationship between techniques is built upon formally in SSADM where cross-checking different types of models against each other is promoted as part of the iterative quality assurance procedures at various stages of the methodology (Goodland and Slater 1995). A similar approach has been proposed for the joint development of entity-relationship diagrams and DFDs (Batini *et al.* 1992).

While recognizing the previous comment concerning the inherent difficulty of describing processes in natural language, a brief description of each of the main stages, represented by the ovals in figure 3, is provided below. The remainder of the paper then provides a worked example to illustrate the application of the pentanalysis technique.

2.2.1. *Task analysis*:   As outlined in §1.1, the task analysis component represented in figure 3 primarily concerns the generation of an activity list from the data captured and is expected to proceed in a fashion common to most task analysis methods. While the one heuristic (§1.1) is still expected to be applied, which will allow the same activity list to be used by other task analysis techniques such as TAKD, the simple selection of actions and objects if replaced by the activity list to pentanalysis table conversion described below (§2.2.3). Iteration is catered for between this conversion and the task analysis that produces the activity list.

2.2.2. *Data flow analysis*:   The data flow analysis proceeds as outlined in §1.2. As the pentanalysis technique is designed to help in the process of analysing the system, it is envisaged that the initial data flow analysis will produce a high level set of DFDs (e.g. levels 0 and 1) and that lower level DFDs will be produced during the pentanalysis technique. The technique's process of integrating the task analysis representation and the DFDs may prompt the analysts to re-evaluate either the current level of abstraction or stimulate the production of lower level DFDs.

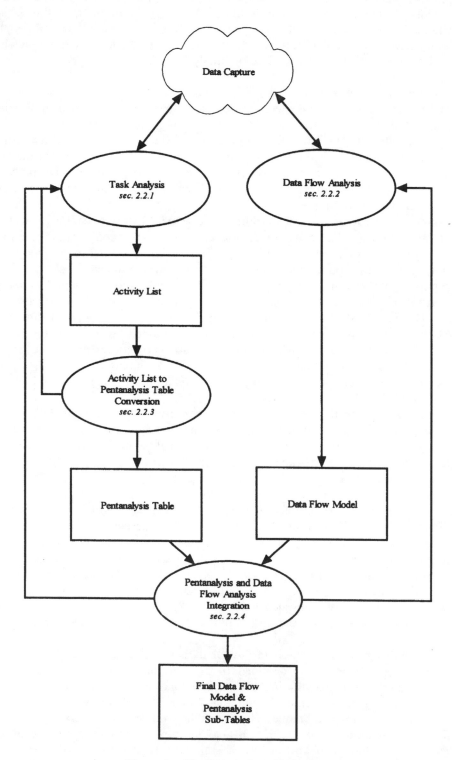

Figure 3.   The Pentanalysis Technique.

2.2.3. *Activity list to pentanalysis table conversion*: The activity list describes a sequence of actions performed by entities in the system. The DFDs describe data flow, data stores and transformations that take place in the system. To produce an integrated activity list and DFD description, the pentanalysis technique first redescribes the activity list in a novel format, the pentanalysis table, which is designed to be compatible with the structure of DFDs.

The pentanalysis table consists of five main columns based on a conceptual model of a data flow. A data flow specifies the flow of data from one data flow element (terminator, process or data store) to another element. The data flow arrow consists of three parts: the element it comes from; the link; and the element it goes to (the arrow head). Figure 4 shows these five components and gives them mnemonic labels. The pentanalysis technique involves taking the words in each line from an activity list and breaking the line down into these five components. This conversion from activity list to pentanalysis table requires two things:

(1) Each of the five pentanalysis table components must have at least one word from each activity list line assigned to it.
(2) All words in an activity list line should be assigned to the pentanalysis table.

Where these conditions are not met then the analyst should return to the initial task analysis stage and revise the activity list. This may involve collecting further data or reinterpreting existing data. All task analyses, irrespective of the method used, involve such iteration, not least, because of the subjective nature of task data (Diaper 1989b).

As will be illustrated in the worked example, the first of these requirements is usually not met because the analyst has omitted an object, often inconsistently with other activity list lines. For example, if an activity list line involves typing on a PC then this assumes a keyboard, which it is easy for the analyst to overlook and not specify in the activity list line.

The second of these requirements arose from the empirical work that led to the development of the pentanalysis technique. Rigorously considering assignment of every word from an activity list line provided analytical leverage in that it forced the authors to re-examine carefully each activity list line that did not meet this requirement.

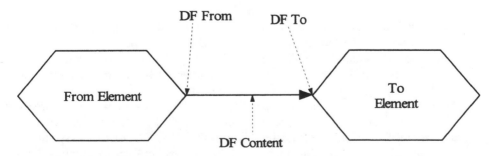

Figure 4. Conceptual relationship between data flow and a pentanalysis line.

2.2.3.1. *An activity list single line example*:   To illustrate how activity list lines are converted into their pentanalysis table format, the following activity list line:

```
The user types the command on the keyboard
```

would be divided into:

```
From Element: The user
DFD From:     types
DFD Content:  the command
DFD To:       on
To Element:   the keyboard
```

Order within the activity list line is not important so that had the line instead said:

```
The command is typed on the keyboard by the user
```

then a semantically identical pentanalysis results:

```
From Element: the user
DFD From:     is typed . . . by
DFD Content:  The command
DFD To:       on
To Element:   the keyboard
```

It would be perverse for an analyst to deliberately mix the grammar of activity list line sentences and, in general, active affirmative sentences are preferred to the passive form of the second example. There are instances in the worked example, however, where the pentanalysis does force order changes from those of the natural language activity list. Indeed, it was these cases that were most helpful to the authors as much of the analysis method involves considering the 'From Element' for each line.

The original, serendipitous discovery that every word in an activity list line could be redescribed as a pentanalysis line turns out to provide an extremely powerful way of checking that the activity list line itself is both well phrased and complete.

The output from the activity list to pentanalysis table conversion process is a six-column table. The first column contains the activity list line number and the rest are based on the five-component conceptual model of a DFD illustrated in figure 4.

2.2.4. *Pentanalysis task and data flow analyses integration*:   While the pentanalysis table is based on a conceptual model of DFDs, the pentanalysis technique must provide a means by which the task analysis data represented in the pentanalysis table can be related to the data represented by the initial hierarchical set of DFDs.

The pentanalysis technique uses a series of assignments in which the analyst makes decisions about the relationship between task activities described as pentanalysis table lines and as data flow elements. The result of these assignments is a series of sub-tables that document the relationship between pentanalysis table lines and DFD elements.

2.2.4.1. *Pentanalysis sub-tables*: Generating the initial pentanalysis table from the activity list is performed in a single step as described in §2.2.3. Activities, represented as lines in the pentanalysis table, must be allocated to elements of a DFD. In complex systems, data flow analysis will actually produce a number of individual DFDs organized in a hierarchy. Allocating task activities to elements in the DFD is an iterative process that starts with the initial pentanalysis table and DFD 0 and proceeds by identifying subsets of the pentanalysis table that are allocated to DFDs lower in the hierarchy.

Each pentanalysis line must be assigned to one, and only one, DFD element at a particular level of a data flow analysis. The heuristic the authors' found most useful was to consider the 'Form Element' in each pentanalysis line as the first basis of assignment. The analyst must employ craft skill here as the 'From Element' contains only a fragment of a natural language description of a task and this is unlikely to match the label given to the data flow element. The analyst must have a clear conception of the functionality of each data flow element.

Figure 5 shows three consecutive lines from an activity list represented in pentanalysis table format.

Figure 6 shows a fragment of a DFD and where the analyst has assigned each of these three lines to the DFD. Line 006 has been assigned to the data flow from 'Data Store' to 'Main Process' (N.B. the analyst has had to infer that 'The computer system' in the pentanalysis table refers to this particular data store in the DFD). Similarly, line 008 has been assigned to a data flow, in this case from 'Main Process'. In contrast, line 7 has been assigned to 'Main Process' itself because it is a process within 'Main Process'. Were the analyst to be interested in the cognitive modelling of the user, within a DFD format, then line 007 would be a candidate for expansion at a lower level of data flow analysis, i.e. 'reads' could be expanded to differentiate

Line No.	From Element	DFD From	DFD Content	DFD To	To Element
006	The computer system	displays	the data	on	the VDU
007	The user	reads	the data	displayed on	the VDU
008	The user	types	<return>	on	the keyboard

Figure 5.   Three activity list lines represented in pentanalysis table format.

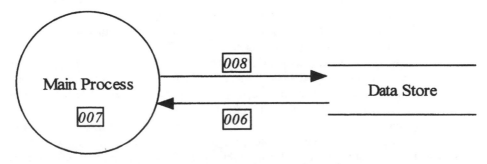

Figure 6.   DFD fragment showing pentanalysis line assignment.

separate cognitive states such as goals, perception, decision mechanisms, long- and short-term memory, etc. Such a reinterpretation of 'reads' would in such a case be similar to the inference the analyst made to line 006 with respect to 'The computer system' referring to a particular data store.

While the authors developed a number of heuristics when inventing and exploring the pentanalysis technique, many of these are clearly not a part of the technique itself but rather the sort of craft skill that develops over time. There is, however, one aspect of the above assignment process of pentanalysis table lines to DFDs that should be part of the pentanalysis technique: the priority of assignment.

Pentanalysis table lines can either be assigned to data flows or to other DFD elements (processes, data stores or terminators). The end result of the assignment process is a set of sub-tables, one table for each element in the DFD. These sub-tables contain the assigned pentanalysis table lines. The priority of assignment is such that lines should, whenever possible, be assigned to data flows rather than to other DFD elements. Indeed, pentanalysis table lines that are assigned to DFD processes are candidates for subsequent decomposition into lower level DFDs where many would be assigned to data flows. When further DFD decomposition is undertaken, an important class of data flows are those that cross the boundary of the part of the DFD that is being expanded by the decomposition because these data flows must, within all traditional data flow analyses, be preserved unaltered ('information flow continuity'; Pressman 1987).

3. Worked example

Having already noted the difficulty of describing methods in a context-independent manner, the following section contains a worked example of the application of the pentanalysis technique. This section also contains some examples of the types of problem that were encountered and the way they were solved. The reason for including this section is two-fold. First, it is intended to be educational by offering advice and examples of the typical 'warts and all' detailed work of the analyst. Second, this paper describes a new method and this section shows that the authors have not cheated by their choices of application domain or task selection. Thus this section is a critical part of the demonstration of the validity and usability of the new method.

3.1. *Background and initial analysis*

The application domain involves a video rental store, which now appears to be a classic exemplar for software engineering method descriptions (Downs *et al.* 1988) and one task has been chosen, that of a customer hiring a video. The scenario is presented in appendix 1 and is the same as that used by Diaper (1997). The task analysis activity list (appendix 2) is also the same as the one in Diaper (1997), although it has been modified during the process of doing the pentanalysis and data flow analysis. One reason for choosing this example is that the activity list was readily available and had already been subjected, at least in part, to intense scrutiny. The activity list is actually a scenario (Carey *et al.* 1991, Clarke 1991, Young and Barnard 1991) and is a tidy example in that it contains no errors. It is, of course, assumed that many activity lists, based on observational data or scenarios, would be analysed in a real project. Such a project would also have to look at a variety of different tasks. For example, at least once a day there must be a 'cashing up' exercise where the till receipts and monies are matched with what is in the loans book.

The modifications made to the activity list before the pentanalysis began were necessary because some lines were incomplete and inconsistent as Diaper's previous work on this activity list had only subjected some lines to analysis. The most obvious inconsistency was the omission of the location of a computer-generated display; this was defaulted, in all cases to 'on VDU'. More interestingly, once the pentanalysis was attempted a similar type of error was found in those lines that modelled human cognitive activity. For example, lines 9 and 26 originally stated that:

```
09    staff remembers customer number
26    staff remembers video number from memory
```

This was modified by the addition to both lines of a terminal 'to himself', which functioned in the same way as the 'on vdu' addition. It was done to fit the pentanalysis, which otherwise would either have no 'DFD To' and 'To Element' for line 9 and would have had to assign 'from' and 'memory' in a manner thought to be confusing in the case of line 26. In parentheses the word 'staff' was added to remind the analysts of the identity of 'himself' as the description represents a reflexive process.

For present purposes it was not considered necessary to undertake detailed user cognitive modelling as this was not ever expected to be part of the DFDs to be produced. In some applications, however, where user modelling is a primary concern, such activity list lines might be further expanded, perhaps to represent the type of human memory accessed, for example, long- or short-term memory, or semantic versus episodic memory.

Figure 7 shows the DFD 0, containing a single process, 'Rent Video' and a terminator, the customer. It took a few minutes to agree and is uncontroversial.

The DFD 1 in figure 8 contains two processes, 'Process Video' and 'Get Video'. 'Process Video' deals with accepting the customer's request and checking their details and 'Get Video' handles the retrieval of the actual video cassette (held in store) and the payment issues. Figure 8 was arrived at by a standard data flow analysis. As the purpose of the work reported in this paper is to demonstrate the 'pentanalysis technique, this initial DFD is a 'first-cut' version based on a brief reading of the scenario. The naming of processes and the exact content of data flows was not considered at this stage and a number of alternatives to this particular level 1 DFD were discussed.

It is worth noting that at this stage the authors decided that they would not normally continue with decomposition below that of DFD 1. These 0 and 1 level

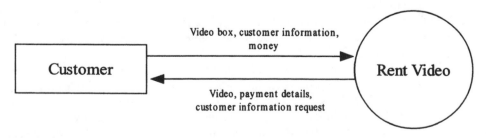

Figure 7.   DFD 0.

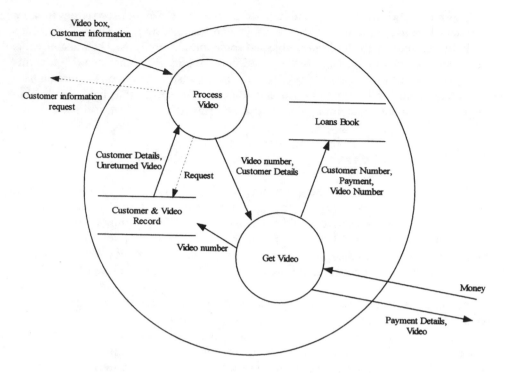

Figure 8.   Initial DFD 1.

DFDs were constructed by considering the background to the project (appendix 1) and generating informal task scenarios, as is usual when constructing DFDs (§1.2 and 1.3). The activity list from Diaper (1997) was occasionally referred to as a source of such task scenarios but was not the major focus of discussion during the initial DFD construction exercise. Both these DFDs were constructed prior to applying the pentanalysis technique, indeed, before the pentanalysis technique was discovered.

Control information (§1.2) has been included and is represented by the dashed lines. The data store 'Loans Book' is included, even though it only possesses input from a single process, so as to act as a mnemonic to the analysts as this data store must take other data flows in a broader analysis than the one used here to illustrate the pentanalysis technique.

3.2. *Applying the pentanalysis technique*

When the pentanalysis starts there is a single pentanalysis table (appendix 3) obtained from the original activity list (appendix 2). Therefore, all pentanalysis lines to be assigned in DFD 0 are contained in this table. As will be seen, this is not the case for lower level DFDs.

In DFD 0 (figure 7), there is only one terminator, 'Customer', for the main 'Rent Video' process with two data flows (in and out). The appearance of single data flows in and out of the 'Rent Video' process is merely a diagrammatic convenience and in larger analyses these would usually be represented by different data flow. If the data flow analysis and task analysis are consistent, it should be possible to allocate each pentanalysis table line to one of these elements.

The pentanalysis lines 1, 3 and 42 in table 1 can be allocated to the data flow from 'Customer' to 'Rent Video'. In the scenario these are points at which the customer interacts with the video shop, i.e. request and pay for a video. In a similar way, lines 2, 41 and 43 can be allocated to the data flow from 'Rent Video' to 'Customer' as they are the points at which the system interacts with the customer, i.e. provide the video and request payment. The remaining pentanalysis lines are all allocated to the DFD 'Rent Video' process. Figure 9 shows the two level 0 sub-tables for these data flows. They illustrate the integration of the data flow and task analyses, which is the purpose of the pentanalysis technique.

In this example, there is no ambiguity about activities being performed in the 'Customer' terminator. The 'Customer' sub-table is empty. It can be imagined that the activity list may have included those tasks performed when the customer selected the video from a shelf in the store. While these tasks are part of a broader 'rent video' process, the appearance of 'Customer' in DFD 0 suggests that the tasks would be allocated to the Customer. Customer is external in the current interpretation of the system, therefore the analysts would have to re-evaluate the DFD boundary with respect to customer in a broader DFD analysis.

A pentanalysis was attempted on DFD 1 in figure 8 using the sub-tables produced for DFD 0. The relevant sub-tables are 'Rent Video', 'Customer'→'Rent Video' and 'Rent Video'→'Customer'.

The data flows described in the level 0 sub-tables 'Customer'→'Rent Video' and 'Rent Video'→'Customer' are those that cross the boundary of the process 'Rent Video' that is to be expanded at level 1. Information flow continuity (§2.2.4.1) must be maintained, which means that only the six data flows shown in figure 9 may cross the 'Rent Video' process boundary. These six will only be different at level 1 from level 0 in that the process within the 'Rent Video' process will be specified at level 1. Lines 1 and 3 can be assigned as data flows into 'Process Video' (sub-table 'Input'→'Process Video') and line 2 can be assigned to data flows out of 'Process Video' (sub-table 'Process Video'→'Output'). In a similar way, lines 41 – 43 can be assigned to input or output sub-tables for 'Get Video'.

Pentanalysis lines 4 to 21 are part of the interaction between the 'Process Video' process and the 'Customer and Video Record' data store. The 'Customer and Video Record' data store is a complex data structure that may be decomposed into two or more data stores. For illustrative purposes it has not been decomposed

'Customer' → 'Rent Video' I/O Data Flow

No.	From Element	DF From	DF Content	DF To	To Element
1	customer	places	empty video box	on	counter
3	customer	tells	customer number	to	staff
42	customer	pays	correct amount	to	staff

'Rent Video' → 'Customer' I/O Data Flow

No.	From Element	DF From	DF Content	DF To	To Element
2	staff	asks	customer number	from	customer
41	staff	requests	payment	from	customer
43	counter	from	plain video box	takes	customer

Figure 9. Level 0 sub-tables of the data flows.

here. Assigning these lines to data flow elements demonstrates one of the major issues raised by integrating DFDs and activity lists below DFD 0: what activity constitutes a data flow? For instance, in this example many of the activities involved the staff member selecting a menu option or acknowledging an action. Technically, these actions frequently trigger tasks by the computer that are not documented in the activity list, for example, saving or committing information in a database. Also many of these actions lead to requests for information from a database, e.g. stating the primary key of a required record, but traditionally a DFD does not show requests for information. This led the authors to represent requests for information in the DFDs (dashed lines). This is consistent with many proposed data flow analysis representations and is particularly useful in this analysis.

The flow of data is clearly illustrated in the pattern that emerges from lines 4 to 21. Those activities that involve the computer's response to the user's actions (e.g. lines 8, 12, 15 and 19) were assigned as 'data flows into a process'. Likewise, those activities that involve the user's response to the computer's actions were assigned as data flows out of a process. This produces an alternating sequence of assignments of one or more processes always followed by a data flow 'To' and then a data flow 'From'. These cycles are indicated by the double lines in figure 10.

Line 21 is interesting because it represents a data flow from 'Process Video' to 'Get Video'. Line 20 could be the data flow but line 21 appears more appropriate, i.e. the staff member reads the video number and then uses it to retrieve the video.

It was initially assumed that lines 22 to 49 would be part of the single process 'Get Video'. Specifically, although lines 22–23 are part of 'Get Video', i.e. retrieving the video cassette from storage, it is not obvious where lines 24–25 should be assigned. Line 24 is a data flow, i.e. the video number is read and will be used to update the loans book (line 25) and is also related to lines 34–35 (updating payments in the loan book). At this point in the analysis, it was decided that lines 24–25 and 34–37 (see the fragment of the pentanalysis sub-table in figure 11) were a single process, although taking place at different points in the activity list, and that this

No.	Cycle	From Element	DF From	DF Content	DF To	To Element
4	Process	computer	displays	main menu	on	VDU
5	"	VDU	on	menu option customer record from main	reads	staff
6	"	staff	selects	menu option customer record from main	via	keyboard
7	To	keyboard	input	menu option customer record from main	accepts	computer
8	From	computer	prompts	customer number	on	VDU
9	Process	staff	remembers	customer number	to	himself
10	"	staff	types	customer number	on	keyboard
11	To	keyboard	input	customer number	accepts	computer
12	From	computer	displays	customer record	on	VDU
13	Process	staff	selects	menu option unreturned videos from	via	keyboard
14	To	keyboard	input	unreturned videos option from customer	accepts	computer
15	From	computer	displays	unreturned videos data	on	VDU
16	Process	VDU	on	customer record for unreturned videos data	checks	staff
17	"	staff	selects	menu option unreturned videos data	via	keyboard
18	To	keyboard	input	unreturned videos data close option	accepts	computer
19	From	computer	displays	customer record	on	VDU
20	Process	staff	orientates	empty video box	on	counter
21	To	video box	on	video number	reads	staff

Figure 10. The repeating cycle of assignments.

process was as important as dealing with the customer's request ('Process Video') or retrieving the video from the store ('Get Video').

Based on the above observation and the requirement that processes should be 'even-sized' and 'well-distinguished', the decision was made to split the 'Get Video' process in figure 8 into two processes: (i) 'Get Video', i.e. retrieving the video from storage and giving it to the customer, and (ii) 'Payment', i.e. requesting and receiving payment from the customer. In this new model, the checking for overdue videos and recording of the video request is performed in 'Process Video', the retrieval of the video cassette and supplying it to the customer is performed in 'Get Video', and the calculation, request and recording of payments is performed in 'Payment'. The new DFD is shown in figure 12.

One of the main reasons for preferring the three process level 1 DFD is that the two activities, retrieving the video and handling the payment, could be performed by different members of staff. Another reason is that the example scenario has a paper-based loan book, which is destined for computerization in the new system (appendix 1). The pentanalysis process has helped to identify this possibility by linking the physical steps taken in the scenario with an abstract description of the process. The problem of abstracting from an existing set of activities to a new system proposal is one of the criticisms of task analysis made by Benyon (1992a,b), although Diaper and Addison (1992b) claim that one of the advantages of task analysis is that it is device-dependent whereas system development tools (e.g. DFDs) erroneously strive for device independence. The method proposed by this paper attempts to reconcile these differences by proposing a systematic method of abstracting the activities captured by task analysis and representing them in a DFD. This is done by using a series of assignments in which the analyst makes decisions about the relationship between activities and data flow elements.

Although data flow analysis, and different analysts, may have produced this conclusion without using the pentanalysis techniques, the result of using it in this case has produced a 'better' DFD. The authors accept that this result could have been reached without using the technique, but the procedure that has been described must take place between a group of analysts working on the construction of a DFD, i.e. discussing alternatives and working towards a consensus. The use of the pentanalysis technique has highlighted these decisions and it has forced us to justify

Process 'Get Video'

No.	From Element	DF From	DF Content	DF To	To Element
22	staff	gets	video tape with video number	from	storage area
23	staff	places	video tape	on	counter
24	video box	on	video number	reads	staff
25	staff	copies	video number from video tape	in	loans book

Process 'Payment'

No.	From Element	DF From	DF Content	DF To	To Element
34	VDU	on	customer number from customer record	reads	staff
35	staff	writes	customer number	in	loans book
36	staff	calculates	payment value	to	himself (staff)
37	staff	writes	payment value	in	loans book

Figure 11.   Example pentanalysis sub-tables.

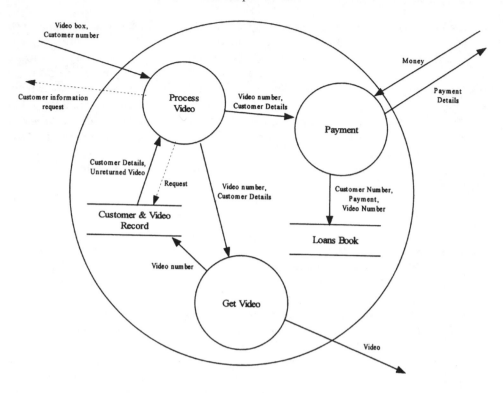

Figure 12.   Final DFD 1.

each conclusion in terms of the observed tasks. It is believed that the pentanalysis technique will provoke questions about any large data flow analysis because a complex problem has many different viewpoints and a single 'correct' DFD analysis is not possible (§1.2).

The remaining lines can be allocated in a similar manner to those described above. Lines 24 and 34, reading the video number and customer number, are data flows between 'Process Video' and 'Payment' leading to the staff member calculating the video charge in lines 34–37. Lines 25 and 35, updating the loan book, are data flows to the 'Loans Book' data store. Lines 40–42 deal with charging the customer and receiving payment. Note that line 43, handing the customer the video, takes place in 'Get Video' although there is no explicit data flow from 'Payment' where lines 40–42 occur. Again, the rationale for this is that these two processes could be performed by separate members of staff.

Appendix 4 shows the complete set of sub-tables for DFD 1 and figure 13 shows the final DFD 1 with pentanalysis table lines superimposed on it.

4.   Discussion and conclusions

The purpose of this paper was first and foremost to provide a description and demonstration of the pentanalysis technique as a method of integrating an anthropocentric requirements method based on task analysis with a computer centric design method using data flow analysis. That this was demonstrated to be possible should, in itself, be of interest to those in the HCI and software engineering

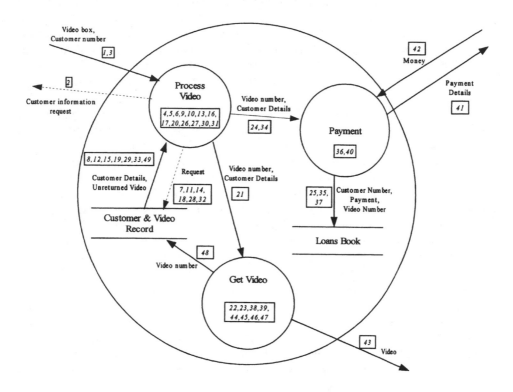

Figure 13. Final DFD 1 and pentanalysis results.

communities concerned with the early stages of the software life-cycle because it allows these particular human and computer centred perspectives to be related to each other in a manner that has never before been possible.

A second issue is the possibility of a synergistic interaction whereby the DFD analysis improves the task analysis and vice versa. Improvements to the task analysis and DFD have been demonstrated to occur: in the DFD when processes are re-interpreted; and in task analysis when activity lines that were missed in the first generation activity list are identified. A second advantage to the task analysis, alluded to but not demonstrated, is that the pentanalysis technique is likely to allow the task analyst to target the acquisition of new task data that is relevant to the needs of the data flow analysis. Given the common problem of data overload often associated with task analysis, this may provide a crucial advantage that may make task analysis more acceptable and common in the commercial software industry by significantly reducing analysis costs. Dillon *et al.* (1993), for example, provides survey evidence of the poor market penetration of task analysis within the European software industry.

While the process of integrating activity lists and DFDs is inherently subjective, the authors feel that they gained very considerable insight into both the logical nature of the processes represented in the DFDs and about the nature of the data flows themselves. There is no doubt that the authors feel extremely confident about their understanding of the final DFD 1 and this is due to the level of scrutiny forced on them by the detail contained in the task analysis. Working collaboratively many

DFD options were investigated and it was found that pentanalysis not only allowed a clearer focus on the issues that were needed to be addressed than hitherto, but provided the authors with a language and explicit rationale for the decisions that the authors had made about the DFDs. It is believed that the work described in this paper will scale up to a complete project and that the benefits of the pentanalysis technique to the production and validation of DFDs would increase with a larger project.

Obviously there is a need to undertake further research in this field using different application examples. In particular, the issue of scalability is of concern, although in large projects it is not expected that every DFD will have a corresponding pentanalysis. The amount of effort required to manually generate the pentanalysis tables and associate them with the DFDs is obviously large, although no more so than with most other task analysis methods, and it will therefore be desirable to automate the technique. It should be fairly easy to produce a simple software toolkit that will allow an activity list to be represented as an initial pentanalysis table by the analyst using 'cut and paste' type operations. It would then be mechanically trivial for the analyst with such a toolkit to associate the pentanalysis sub-tables and the DFDs, thus freeing the analyst to concentrate on the critical analysis decisions. Ultimately it might be expected that such a software tool may be incorporated into Computer Assisted Software Engineering (CASE) tools (e.g. Spurr and Layzell 1990) that already support data flow analysis and the generation of DFDs. Such CASE tools are already fairly common in the software industry (Baba 1990, for example evaluates six SSADM case tools) and the cost overheads of replacing the current, rather casual, approach to modelling the world for the purposes of producing DFDs would be trivial within such CASE tools which already contain far larger and more exotic types of data than those of the pentanalysis sub-tables, for example, the logical Data Dictionary Entries (DDEs) that are always supposed to be associated with the production of DFDs.

In the longer term, the integration of the pentanalysis technique with Diaper's work on task to entity modelling should provide a means for HCI to contribute to two of the three main system views of SSADM and therefore allow an HCI-orientated method to have a major contribution to the early stages of the software life cycle that does get incorporated in the subsequent design and later stages in the software development process.

References

ANNETT, J. and DUNCAN, K. D. 1967, Task analysis and training design, *Occupational Psychology*, **41**, 211–221.

ANNETT, J., DUNCAN, K. D., STAMMERS, R. B. and GRAY, M. J. 1971, Task analysis, Training Information Paper No. 6, HMSO, London. Reprinted 1980, Training Information Paper Series, Manpower Services Commission, London.

ASHWORTH, C. and GOODLAND, M. 1990, *SSADM: A Practical Approach* (UK: McGraw-Hill).

BABA, I. 1990, A standard for SSADM CASE support, in K. Spurr and P. Layzell (eds), *Case on Trial* (New York: Wiley).

BATINI, C., CERI, S. and NAVATHE, S. B. 1992, 'Conceptual database design' (Benjamin and Cummings).

BENCH-CAPON, T. and McENRY, A. 1989a, People interact through computers not with them, *Interacting with Computers*, **1**, 31–38.

BENCH-CAPON, T. and McENRY, A. 1989b, Modelling devices and modelling speakers, *Interacting with Computers*, **1**, 220–224.

BENYON, D. 1992a, The role of task analysis in systems design, *Interacting with Computers*, **4**, 102–123.

BENYON, D. 1992b, Task analysis and system design: the discipline of data, *Interacting with Computers*, **4**, 246–259.

BOEHM, B. W. 1975, The high cost of software, in E. Horowitz (ed.), *Practical Strategies for Developing Large Software Systems* (Reading, MA: Addison-Wesley).

CAMERON, J. R. 1983, *JSP & JSD: The Jackson Approach to Software Development* (Los Angeles, USA: IEEE Computer Society Press).

CAREY, T., McKERLIE, D., BUBIE, W. and WILSON, J. 1991, Communicating human factors expertise through usability design rationales and scenarios, in D. Diaper and N. Hammond (eds), *People and Computers VI* (Cambridge: Cambridge University Press), 117–132.

CLARKE, L. 1991, The use of scenarios by user interface designers, in D. Diaper and N. Hammond (eds), *People and Computers VI* (Cambridge: Cambridge University Press), 103–116.

COTTS, G. 1987, *Structured systems analysis and design methodology* (London: Paradigm).

DAVIS, A. M. 1990, *Software Requirements: Analysis and Design* (Englewood Cliffs, NJ: Prentice-Hall).

DEMARCO, T. 1979, *Structured Analysis and System Specification* (Englewood Cliffs, NJ: Prentice-Hall).

DIAPER, D. 1989a, Task analysis for knowledge descriptions (TAKD): The method and an example, in D. Diaper (ed.), *Task Analysis for Human-Computer Interaction* (Chichester: Ellis Horwood), 108–159.

DIAPER, D. 1989b, Task observation for human-computer interaction, in D. Diaper (ed.), *Task Analysis for Human-Computer Interaction* (Chichester: Ellis Horwood), 210–237.

DIAPER, D. 1989c, Designing expert systems: from Dan to Beersheba, in D. Diaper (ed.), *Knowledge Elicitation: Principles, Techniques and Applications* (Chichester: Ellis Horwood), 15–46.

DIAPER, D. 1989d, *Task Analysis for Human-Computer Interaction* (Chichester: Ellis Horwood).

DIAPER, D. 1990a, Simulation: a stepping stone between requirements and design, in A. Life, C. Narborough-Hall and W. Hamilton (eds), *Simulation and the User Interface* (London: Taylor & Francis), 59–72.

DIAPER, D. 1990b, Analysing focused interview data with task analysis for knowledge descriptions (TAKD), in D. Diaper, D. Gilmore, G. Cockton and B. Shackel (eds), *Human-Computer Interaction: Interact'90* (Amsterdam: North Holland), 277–282.

DIAPER, D. 1997, Integrating human-computer interaction and software engineering requirements analysis: a demonstration of task analysis supporting entity modeling, *SIGCHI Bulletin*, **29**, 1.

DIAPER, D. and ADDISON, M. 1991, User modelling: The task oriented modelling (TOM) approach to the designer's model, in D. Diaper and N. Hammond (eds), *People and Computers VI* (Cambridge: Cambridge University Press, 387–402.

DIAPER, D. and ADDISON, M. 1992a, HCI: The search for solutions, in A. F. Monk, D. Diaper and M. D. Harrison (eds), *People and Computer VII* (Cambridge: Cambridge University Press), 493–495.

DIAPER, D. and ADDISON, M. 1992b, Task analysis and systems analysis for software engineering, *Interacting with Computers*, **4**(1), 124–139.

DIAPER, D. and JOHNSON, P. 1989, Task analysis for knowledge descriptions: theory and application in training, in J. Long and A. Whitefield (eds) *Cognitive Ergonomics and Human-Computer Interaction* (Cambridge: Cambridge University Press), 191–224.

DILLON, A., SWEENEY, M. and MAGUIRE, M. 1993, A survey of usability engineering within the European IT industry—current practice and needs, in J. L. Alty, D. Diaper and S. Guest, *People and Computers VIII* (Cambridge: Cambridge University Press), 81–94.

DOWNS, E., CLARE, P. and COE, I. 1988, *Structured Systems Analysis and Design Method: Application and Context* (Englewood Cliffs, NJ: Prentice-Hall).

EVA, M. 1994, *SSADM Version 4: A User's Guide*, (2nd edn) (New York: McGraw-Hill).

GIKAS, S., JOHNSON, P. and REEVES, S. 1990, Formal framework for task oriented modelling of devices, Department of Computer Science, Technical Report, Queen Mary and Westfield College.

GOODLAND, M. and SLATER, C. 1995, *SSADM V4: A Pratical Approach* (New York: McGraw-Hill).

HARES, J. 1990, *SSADM for the Advanced Practitioner* (New York: Wiley).

JANE, C. and SARSON, T. 1979, *Structured Systems Analysis: Tools and Techniques* (Englewood Cliffs, NJ: Prentice-Hall).

JOHNSON, P. 1985, Towards a task model of messaging: an example of the application of TAKD to user interface design, in P. Johnson and S. Cook (eds), *People and Computers: Designing the Interface* (Cambridge: Cambridge University Press), 46–62.

JOHNSON, P., DIAPER, D. and LONG, J. 1984a, Syllabi for training in information technology, in E. Megaw (ed.), *Contemporary Ergonomics 1984* (London: Taylor & Francis), 92–97.

JOHNSON, P., DIAPER, D. and LONG, J. 1984b, Tasks, skills and knowledge: task analysis for knowledge based descriptions, in B. Shackel (ed.), *Interact'84—First IFIP conference on Human-Computer Interaction* (Amsterdam: Elsevier), 23–27.

JOHNSON, P., DIAPER, D. and LONG, J. 1985, Task analysis in interactive systems design and evaluation, in G. Johannsen, G. Mancini and L. Martensson (eds), *Analysis, Design and Evaluation of Man-Machine Systems* (Oxford: Pergamon Press), 123–127.

LIM, K. Y. and LONG, J. 1994, *The MUSE Method for Usability Engineering* (Cambridge: Cambridge University Press).

LONG, J. 1986, People and computers: designing for usability, in M. Harrison and A. Monk (eds), *People and Computers: Designing for Usability* (Cambridge: Cambridge University Press), 3–23.

LONG, J. 1989, Cognitive ergonomics and human-computer interaction: an introduction, in J. Long and A. Whitefield (eds), *Cognitive Ergonomics and Human-Computer Interaction* (Cambridge: Cambridge University Press), 4–34.

LONG, J. and DOWELL, J. 1989, Conceptions of the discipline of HCI: craft, applied science, and engineering, in A. Sutcliffe and L. Macaulay (eds), *People and Computers V* (Cambridge: Cambridge University Press), 9–34.

PAYNE, S. and GREEN, T. 1986, Task-action grammars: a model of the mental representation of tasks, *Human-Computer Interaction*, **2**, 93–133.

PAYNE, S. and GREEN, T. 1989, Task-action grammars: the model and its developments, in D. Diaper (ed.), *Task Analysis for Human-Computer Interaction* (Chichester: Ellis Horwood), 75–107.

PRESSMAN R. S. 1987, *Software Engineering: A Practitioner's Approach*, 2nd edn (New York: McGraw-Hill).

PRESSMAN, R. S. 1994, *Software Engineering: A Practitioner's Approach*, 3rd edn, European adaptation (New York: McGraw-Hill).

SOMMERVILLE, I. 1989, *Software Engineering*, 3rd edn (Reading, MA: Addison-Wesley).

SOMMERVILLE, I. 1996, *Software Engineering*, 5th edn (Reading, MA: Addison-Wesley).

SPURR, K. and LAYZELL, P. 1990, *Case on Trial* (New York: Wiley).

TAYLOR, F. W. 1911, *Principles of Scientific Management* (New York: Harper and Row).

YOUNG, R. and BARNARD, P. 1991, Signature and paradigm tasks: new wrinkles on the scenarios methodology, in D, Diaper and N. Hammond (eds), *People and Computers VI* (Cambridge: Cambridge University Press), 91–102.

YOURDON, E. 1989, *Modern Structured Analysis* (London: Yourdon Press).

Appendix 1. Background task scenario

The demonstration application domain is a UK high street video hire shop. Such shops typically carry a stock of between 500 and 2000 videos that may be hired on a daily basis, generally to a geographically restricted population. In the example chosen for demonstration purposes, the particular video shop used for analysis is characterized as having a mixed technology system.

The shop has a small Personal Computer (PC) that stores records of customer contact details (name, address, telephone and their indentification number) and keeps a record of each video borrowed. In parallel with the PC, the shop keeps a paper record in a loans book of each transaction. All cash details are only recorded in the loans book. The shop in the example is one of a small chain of such shops. The chain's owners wish to upgrade and unify their shop and central office PCs with the long-term intention of providing a real-time, interactive, distributed computing environment that will improve customer service and reduce stock replication by being able to identify where a particular video is in any of the chain's shops on a customer request. This target will be approached in several stages by the chain's management. The first stage is improvement of the in-shop, stand-alone application which has, in any case to be carried out, because of pressure from the company's accountants to remove the loans book part of the system.

The task analysis carried out is part of the early, requirements analysis stage of this project's first stage. The whole, distributed computing project requires a careful analysis of the current system as there would be significant additional costs associated with implementing in-shop, stand-alone PCs if they then require extensive change for the intended distributed services. Such costs involve not only those directly related to software but also human costs such as staff training. It is a management view that the latter should be minimal, which may be a harsh but realistic view of their business.

In the demonstration, a single interaction involving a customer hiring a video is analysed. A full analysis would look at many such transactions. The activity list that was created for the demonstration is a tidy example, it is minimal, contains no errors and has been described in neat prose following the heuristic that an activity list should have no more than one action per line.

Looking at many tasks, the task analysis naturally concentrates on three major component systems:

(1) The current PC system;
(2) The loans book; and
(3) The employee user.

The other component systems identified involve either relations or dealing with physical objects that are candidates for automation, i.e. getting out a video from stock, placing it in a box, etc. Some analysis will be desirable of the employee-customer interaction as it should be a design goal that the new system does not interfere with customer relations. Apart from any sophisticated view about the importance of such social relationships in this market, it is imagined that management policy is to minimize staff levels while not having long queues in the shop (queues lead to loss of video hirings and hence loss of turnover). The overall design approach is to model two systems:

(1) The PC system; and
(2) The loans book system.

The latter will then be redesigned so as to fit PC operations and this redesign will be fitted to the existing PC system. The assumption is that this last stage will produce a new PC system design that may be structurally different from that of the current system. This, of course, is not intended as a full or adequate description of the design process. If SSADM, for example, were being used then its other analysis stages would proceed as usual. Even in some other software development environments, there will be other methods and representations being employed.

Appendix 2. Video shop with current computer activity list

1   customer places on counter empty video box
2   staff asks customer for customer number
3   customer tells customer number to staff
4   computer displays main menu on VDU
5   staff reads menu option customer record from main menu on VDU
6   staff selects menu option customer record from main menu via keyboard
7   computer accepts keyboard input menu option customer record from main menu
8   computer prompts for customer number on VDU
9   staff remembers customer number to himself (staff)
10 staff types customer number on keyboard
11 computer accepts keyboard input customer number
12 computer displays customer record on VDU
13 staff selects menu option unreturned videos from customer record via keyboard
14 computer accepts keyboard input unreturned videos option from customer record
15 computer displays unreturned videos data on VDU
16 staff checks customer record for unreturned videos data on VDU
17 staff selects menu option unreturned videos data via keyboard
18 computer accepts keyboard input unreturned videos data close option
19 computer displays customer record on VDU
20 staff orientates empty video box on counter
21 staff reads video number on video box
22 staff gets video tape with video number from storage area
23 staff places video tape on counter
24 staff reads video number on video tape
25 staff copies video number from video tape in loans book
26 staff remembers video number from memory to himself (staff)
27 staff types video number from memory on keyboard
28 computer accepts keyboard input video number
29 computer displays video title in customer record on VDU
30 staff reads video title on VDU
31 staff types Y for confirm on keyboard
32 computer accepts keyboard input Y for confirm
33 computer displays no error message on VDU
34 staff reads customer number from customer record on VDU
35 staff writes customer number in loans book
36 staff calculates payment value to himself (staff)
37 staff writes payment value in loans book
38 staff gets plain video box from under counter
39 staff places video tape in empty video box
40 staff remembers payment value to himself (staff)
41 staff requests payment from customer
42 customer pays correct amount to staff
43 customer takes plain video box from counter
44 staff selects close option on customer record via keyboard
45 computer accepts keyboard input close option on customer record
46 computer displays confirm message in customer record on VDU
47 staff types Y for confirm on keyboard
48 computer accepts keyboard input Y for confirm
49 computer displays main menu on VDU

Appendix 3.   Pentanalysis table obtained from the activity list

No.	From Element	DF From	DF Content	DF To	To Element
1	customer	places	empty video box	on	counter
2	staff	asks	customer number	from	customer
3	customer	tells	customer number	to	staff
4	computer	displays	main menu	on	VDU
5	VDU	on	menu option customer record from main	reads	staff
6	staff	selects	menu option customer record from main	via	keyboard
7	keyboard	input	menu option customer record from main	accepts	computer
8	computer	prompts for	customer number	on	VDU
9	staff	remembers	customer number	to	himself (staff)
10	staff	types	customer number	on	keyboard
11	keyboard	input	customer number	accepts	computer
12	computer	displays	customer record	on	VDU
13	staff	selects	menu option unreturned videos from	via	keyboard
14	keyboard	input	unreturned videos option from customer	accepts	computer
15	computer	displays	unreturned videos data	on	VDU
16	VDU	on	customer record for unreturned videos	checks	staff
17	staff	selects	menu option unreturned videos data	via	keyboard
18	keyboard	input	unreturned videos data close option	accepts	computer
19	computer	displays	customer record	on	VDU
20	staff	orientates	empty video box	on	counter
21	video box	on	video number	reads	staff
22	staff	gets	video tape with video number	from	storage area
23	staff	places	video tape	on	counter
24	video box	on	video number	reads	staff
25	staff	copies	video number from video tape	in	loans book
26	staff	remembers	video number from memory	to	himself (staff)
27	staff	types	video number from memory	on	keyboard
28	keyboard	input	video number	from	memory
29	computer	displays	video title in customer record	on	VDU
30	VDU	on	video title	reads	staff
31	staff	types	Y for confirm	on	keyboard
32	keyboard	input	Y for confirm	accepts	computer
33	computer	displays	no error message	on	VDU
34	VDU	on	customer number from customer record	reads	staff
35	staff	writes	customer number	in	loans book
36	staff	calculates	payment value	to	himself (staff)
37	staff	writes	payment value	in	loans book
38	under counter	from	plain video box	gets	staff
39	staff	places	video tape	in	plain video box
40	staff	remembers	payment value	to	himself (staff)
41	staff	requests	payment	from	customer
42	customer	pays	correct amount	to	staff
43	counter	from	plain video box	takes	customer
44	staff	selects	close option on customer record	via	keyboard
45	keyboard	input	close option on customer record	accepts	computer
46	computer	displays	confirm message in customer record	on	VDU
47	staff	types	Y for confirm	on	keyboard
48	keyboard	input	Y for confirm	accepts	computer
49	computer	displays	main menu	on	VDU

D. Diaper et al.

Appendix 4. Pentanalysis sub-tables for the final DFD 1

'Process Video → 'Get Video' Data Flow

No.	From Element	DF From	DF Content	DF To	To Element
21	video box	on	video number	reads	staff

'Process Video' → 'Payment' Data Flow

No.	From Element	DF From	DF Content	DF To	To Element
24	video box	on	video number	reads	staff
34	VDU	on	customer number from customer record	reads	staff

'Process Video' → 'Customer and Video Record' Data Flow

No.	From Element	DF From	DF Content	DF To	To Element
7	keyboard	input	menu option customer record from main menu	accepts	computer
11	keyboard	input	customer number	accepts	computer
14	keyboard	input	unreturned videos option from customer	accepts	computer
18	keyboard	input	unreturned videos data close option	accepts	computer
28	keyboard	input	video number	from	memory
32	keyboard	input	Y for confirm	accepts	computer

'Get Video' → 'Customer and Video Record' Data Flow

No.	From Element	DF From	DF Content	DF To	To Element
48	keyboard	input	Y for confirm	accepts	computer

'Customer and Video Record' → 'Process Video' Data Flow

No.	From Element	DF From	DF Content	DF To	To Element
8	computer	prompts for	customer number	on	VDU
12	computer	displays	customer record	on	VDU
15	computer	displays	unreturned videos data	on	VDU
19	computer	displays	customer record	on	VDU
29	computer	displays	video title in customer record	on	VDU
33	computer	displays	no error message	on	VDU
49	computer	displays	main menu	on	VDU

'Payment' → 'Loans Book' Data Flow

No.	From Element	DF From	DF Content	DF To	To Element
25	staff	copies	video number from video tape	in	loans book
35	staff	writes	customer number	in	loans book
37	staff	writes	payment value	in	loans book

Input → 'Process Video' I/O Data Flow

No.	From Element	DF From	DF Content	DF To	To Element
1	customer	places	empty video box	on	counter
3	customer	tells	customer number	to	staff

'Process Video' → Output I/O Data Flow

No.	From Element	DF From	DF Content	DF To	To Element
2	staff	asks	customer number	from	customer

Input → 'Payment' I/O data Flow

No.	From Element	DF From	DF Content	DF To	To Element
42	customer	pays	correct amount	to	staff

'Payment → Output I/O Data Flow

No.	From Element	DF From	DF Content	DF To	To Element
41	staff	requests	payment	from	customer

'Get Video' → Output I/O Data Flow

No.	From Element	DF From	DF Content	DF To	To Element
43	counter	from	plain video box	takes	customer

'Process Video' Process

No.	From Element	DF From	DF Content	DF To	To Element
4	computer	displays	main menu	on	VDU
5	VDU	on	menu option customer record from main	reads	staff
6	staff	selects	menu option customer record from main	via	keyboard
9	staff	remembers	customer number	to	himself (staff)
10	staff	types	customer number	on	keyboard
13	staff	selects	menu option unreturned videos from	via	keyboard
16	VDU	on	customer record for unreturned videos data	checks	staff
17	staff	selects	menu option unreturned videos data	via	keyboard
20	staff	orientates	empty video box	on	counter
26	staff	remembers	video number from memory	to	himself (staff)
27	staff	types	video number from memory	on	keyboard
30	VDU	on	video title	reads	staff
31	staff	types	Y for confirm	on	keyboard

'Get Video' Process

No.	From Element	DF From	DF Content	DF To	To Element
22	staff	gets	video tape with video number	from	storage area
23	staff	places	video tape	on	counter
38	under counter	from	plain video box	gets	staff
39	staff	places	video tape	in	plain video
44	staff	selects	close option on customer record	via	keyboard
45	keyboard	input	close option on customer record	accepts	computer
46	computer	displays	confirm message in customer record	on	VDU
47	staff	types	Y for confirm	on	keyboard

'Payment' Process

No.	From Element	DF From	DF Content	DF To	To
36	staff	calculates	payment value	to	himself
40	staff	remembers	payment value	to	himself

'Customer and Video Record' Data Store

No.	From Element	DF From	DF Content	DF To	To

'Loans Book' Data Store

No.	From Element	DF From	DF Content	DF To	To

CHAPTER 4

Reasoning about tasks, activities and technology to support collaboration

Leon A. Watts* and Andrew F. Monk

Department of Psychology, University of York, York YO1 5DD, UK

Keywords: Collaborative work; Task analysis; Telemedicine; Telecommunication.

An aspect of collaboration described as 'semi-synchronized activity' is discussed as a particular challenge for the task analysis (TA) of collaborative work. TA typically involves the decomposition of work systems into essentially independent component processes between which commodities (information or materials) pass. In collaborative work, people routinely violate the condition of independence by moving seemlessly in and out of synchronization with one another, allowing for both independent and varying levels of conjoint activity. The shift between joint and independent projects is not fixed but managed through more or less explicit awareness of the other people over time. A number of case studies of the effect of communication technologies in telemedical consultation are drawn upon to illustrate the relationship between awareness and synchronization in collaborative work. They show that an analysis of collaborative activity requires a consideration of: (1) the activities constituting work; (2) the interactions between participants required to carry out the activities; (3) who else has access to these activities besides the primary participants in the ongoing work; (4) the contemporaneity of activities; (5) the locations/environments in which the activities are carried out; and (6) the constraints that apply to accessibility and participation within and between these environments. The Comms Usage Diagram is described as a framing notation incorporating these characteristics for a broad, communications-level analysis of collaborative activity. It shows how particular technologies relate to particular phases of work, indexing their effects to collaborative activities in those contexts.

1. Introduction

The value of task analysis has been demonstrated many times since its beginnings in the 1960s (Miller 1962). It has been used in a variety of work contexts and for a number of purposes. It has been applied to problem identification where some work systems failure has occurred, to requirements capture for job or computer system design, and to the design of training for cognitively complex skills (Shepherd 1995). Kirwan and Ainsworth (1992) describe a number of task analysis case studies in contexts ranging from maintenance to control room operations, and from the operation of a large industrial plant to the business of inspecting the quality of welds in high-integrity pressure vessels. In each case, the central concern is to reason about the way in which some work is carried out by people interacting with equipment or with others, under some specified operating constraints, for a particular purpose. To a large extent this comes down to categorization of the operations in work, in that the general activities performed by people in some contexts are classified into

---

* Author for correspondence.

intentions and actions, of greater or lesser cognitive involvement. Categorization has been fundamental to psychology's programme of the understanding and improvement of work since Taylor first set out his *Principles of Scientific Management* (Taylor 1911). Many task analysis techniques rely on systematic categorical structures for hierarchical decomposition (Shepherd 1995), the work system being treated as a series of more or less independent operational units.

1.1. *The pragmatics of task analysis*

The ideas and methods of task analysis have grown up out of the practical concerns of occupational and industrial psychologists to address the problems that they have encountered. As such, the methods associated with task analysis invariably bear the marks of the particular domains in which they were developed. Reviews of task analysis thus tend not to conclude uniformly in favour of particular approaches, but rather indicate the particular strengths or facilities offered by the various methods (Drury *et al.* 1987). In short, they tend to be presented as a series of approaches more or less applicable for various problem areas. Kirwan and Ainsworth (1992), for example, describe 41 different techniques, together with their assessment of the design or operation stage at which they have most to offer.

As well as being adapted for some particular work context, task analysis methods are employed for some well-defined practical purpose, such as the design of training or technology. This practical purpose drives the nature of the analysis. Shepherd (1995), discussing Hierarchical Task Analysis (HTA) as a framework within which particular techniques are applied, states that goals need only be decomposed into sub-goals when there is a practical reason to do so. Let us say that the purpose is design and the proposed technology is reasonably well defined. Some higher level goals will be clearly outside of the area of work to be supported and so can be left as they are. Others that are central to the work to be supported by the technology will be decomposed into sub-goals and sub-sub-goals.

One of the skills of a task analyst is to make a decision about the likely value of investing effort in producing an in-depth TA of some aspect of a work system, against the potential pay-off in terms of improvements to be had. Arriving at an exhaustive description of all the work carried out is not the point; it is to produce a reliable and definitive procedural description of an aspect of the work for some well-defined practical purpose. The variety of available techniques underlines this emphasis on understanding the work in a particular domain for some end, rather than a commitment to particular representational semantics for the theoretical advancement of the understanding of work. However, regardless of the pragmatics of deploying TA techniques, those techniques have some conceptual background that, as we shall see, mitigates against their value in collaborative contexts.

1.2. *The task concept in task analysis*

It is worthwhile to consider here what is meant by the 'task' in task analysis and how the work described in this paper relates to that notion. Preece *et al.* (1994:411) describe tasks as intrinsically purposeful in a way that bare functions are not. A task is comprised of the activities: 'required, used or believed to be necessary to achieve a goal using a particular device', where in their terms a goal is 'the state of a system that the human wishes to achieve' and 'activities are behaviours devoid of any control structure; they do not require thought'. So tasks are set at a higher level than automatic action, although including it ultimately in an instrumental way, set firmly in the realm of the cognitive. Filkes (1982) observed that tasks are not the achievement of a set of goals in isolation but have requirements for maintaining a set of constraints within which plans are formulated. Johnson (1992) emphasizes that the relationship between tasks and their purposes is complex, in that these purposes extend beyond those of the immediate work within which the task is a part to include the intention-base of the persons to carry out the nominal tasks. This is expressed as the sub-goal structure of the individuals concerned. The sub-goal structure conditions selection amongst action alternatives in conjunction with the knowledge base of those concerned. Waern (1989), in a consideration of expertise, noted that some notion of feedback is implicit in the concept of task, as unless the effect of actions performed is evaluated, it is impossible to determine whether or not the task's goal has been achieved. Furthermore, she suggests that task planning includes knowledge about when to initiate the task along with the available methods and tools for carrying out the task.

The task concept then refers to cognitive activity including an *a priori* objective; the desired effect set against a background of a variety of other specifiable objectives and constraints; an understanding of available resources; a plan for achieving the effect including how to exploit the necessary resources; the evidence marking its successful completion; and the time scale of implementing the plan.

1.3. *Task descriptions and work practices*

Studies of various kinds of work with computer systems, in the field of Human-Computer Interaction (HCI), have shown up some difficulties in dealing with work described as a set of tasks in this way. Waern (1989) observed that purposes can be radically underspecified so that a part of an expert's task can be in some sense to create the goal for which their efforts are expended. The notion of purpose in task can very easily become blurred. She goes on to observe that, on implementation, the task concept evolves such that the mismatch between the human-computer divisions of labour anticipated by the user population and the designer become revealed in use leading to 'the user's mental model of the task in the system' evolving and the crucial resource in the task notion becoming utterly transformed. Carroll *et al.* (1991), considering the consequence of introducing new technology to a work system, have described this process as the task-artefact cycle: users get to see previously unanticipated possibilities through the newly introduced tool, leading to changes in their work and a new set of requirements for future design.

Draper (1993), studying individuals interacting with computers, has identified five common kinds of real-user task: initiating an intended command function; verification of the success of the function; exploring alternatives for performing some other task; finding out about the effects of a particular system function; and defence behaviours, such as the 'precautionary save'. On this evidence, he cautions against overly restrictive system specification, a consequence of adherence to a narrow idealized view of user behaviour. Robinson (1993), on a similar theme, implored systems designers to explicitly design in flexibility by avoiding implementations demanding heavily constrained kinds of behaviour. In his terms, one should 'design for unanticipated use'.

A solution to the problems of over-constrained design may be to abstract away from the details of individuals' work. In response to this kind of concern, the rigid, prescriptive, fully specified decomposition of work into its component operations is less common in contemporary HCI. As Preece *et al.* (1994:410) put it: 'Many researchers are tending to focus on the notion of "work" or "activities" rather than the more specific notion of "task". These terms acknowledge the distributed nature of real work situations'. Where work situations are of an overtly collaborative nature, the issue of distribution becomes particularly pressing.

1.4. *Dealing with distributed work situations*
Studies in the ethnographic tradition, developed to serve sociological and anthropological theoretical concerns, have made a concerted effort to account for this 'distributed nature of real work'. They tend to be more concerned with describing what happens in a thorough and unbiased way than in fulfilling some practical brief, although the work of Nardi, for example (Nardi and Miller 1991, Nardi *et al.* 1993), and Greatbatch *et al.* (1995) does show a very pragmatic orientation. Although the centrality of pragmatic purpose to TA is an important difference from these other approaches to the study of work practices, it is by no means the only one. What these sociological approaches have, that traditional task analysis does not, is a general concern with the group rather than the individual and the way work is organized in practice rather than in principle. This is clearly of importance in the design of technology for collaboration.

There is a large body of research concerned with describing technologically supported collaboration under the heading Computer Supported Co-operative Work (CSCW). Much of this has been contributed by computer scientists, sociologists and psychologists, although investigators coming from many other disciplines have played a part (Bowers and Benford 1991, Baecker 1993. Ch. 4 and 5, Rosenburg and Hutchinson 1994). Similarly pragmatic in orientation to TA, the field has evolved subject to analogous pressures. It was recognized quite early on that the apparent concreteness of its subject matter was illusory, in that the huge variety of activities accepted as forms of computer supported co-operative work had little in common. Whereas TA has developed as a wide set of techniques to be selected and adapted to provide abstracted and definitive descriptions of work, the CSCW approach has been to draw out in detail some of the situational aspects that determine the effective operation of the collaborative structures in the groups performing the work (Robinson 1991). Writing in 1991, Bannon and Schmidt noted that CSCW finds its focus in the development of general approaches to understand the business of co-operation first, and thereafter to consider available technologies:

> It has been said that what we have to be concerned about in thinking of computer technology with respect to co-operative work is not the 'support' notion, but first of all ensuring that the computer does not *disrupt* the collaborative activity that is already going on! . . . CSCW should be conceived as an endeavour to understand the nature and characteristics of co-operative work with the objective of designing adequate computer-based technologies (Bannon and Schmidt 1991:4).

The consequence has been an emphasis on revealing the processes of mutual understanding and co-ordination between individuals in particular settings. Local practices and organizational cultures have been found to exert such a strong influence on the utility of particular resources that notionally equivalent work, say group decision support systems (GDSS), are popular and effective in one context and unusable in another.

A contribution to CSCW design from linguistics illustrates both the breadth of disciplinary contribution to the area and also the controversy of introducing general principles as system-enforced rules of work in particular contexts. Winograd and Flores (1986), to take an influential example, used some work in Speech Act Theory to design a computer tool, Co-ordinator, to help to stimulate and moderate interpersonal interactions in a business setting. Co-ordinator embodied a state transition model of conversational structure, so it was assumed, for example, that a 'request' will always be followed by a 'promise', a 'counter' or a 'rejection'. Many investigators in CSCW have questioned this approach because it forces people using it to adopt that model (Bannon 1995). These same investigators, and the present authors, see the value of characterizing the structure in conversation and other collaborations but hold back from actually prescribing it as a part of the design. Again, as pragmatists, it is recognized that the characterization of the work produced in the present analyses of activity is necessarily an approximation of the full repertoire of actions available to collaborators. It would thus be dangerous to build tools that *only* support the method of collaboration that is described. Rather, one should build tools that *at least* support that model of collaboration. This paper uses some of the ideas from ethnographic studies to broaden the analysis of work for pragmatic purposes to provide a usable account of collaboration in practice.

1.5. *Collaboration as a continuum, brokered by interpersonal awareness*

From its inception task analysis has been applied to work systems involving many different people. Collaboration describes more than the inclusion of several agents at work. The difference is to do with the interleaving of, and sensitivity to, the activities of the various parties concerned. Collaboration describes a requirement for communication between, and active co-ordination of, these parties. This contrasts with the kind of work system, often addressed within a Critical Path Analysis framework, say, where participants are conceived as being interdependent only in terms of work flow. People who take part in the kinds of collaborative activity described here are mutually sensitive to one another on a number of levels. This paper sets out to describe some of these sensitivities of material importance to the understanding of collaboration and goes on to indicate how one can represent these for the purpose of reasoning about the design of technological support for collaboration. The term 'collaboration' as used here connotes a class of activity requiring the active and reciprocal involvement of two or more people for the achievement of some joint aim. For each person, conceiving of the job in hand is only possible by considering and allowing for the other people also involved. This view of work sees people as organizing themselves to take some joint action, doing something else individually or with other people, coming together again and so on.

Here, the term synchronization should be understood to mean 'organized in time', rather than purely as simultaneous occurrence, so swimmers are synchronized when they move in harmony with their music and one another, and relay runners are synchronized by carefully alternating their periods of running. Collaborative work is seen as a continuous flux of varying levels of synchronization mediated by mutual awareness. Awareness as a concept in CSCW is multifaceted (McDaniel and Brink 1997). By mutual awareness the authors mean knowledge of one anothers' interpersonal 'status', for example looking towards you, or concentrating hard on some other task (Watts *et al.* 1996).

1.6. *From collaborative tasks to semi-synchronized activity*

This paper takes its unit of analysis at a higher level than 'task' as formerly discussed, in that the sub-cognitive actions of those involved are not dealt with and additional collaborative characteristics are included. Rather than task, the term 'activity' will be used. Not to be confused with the 'behaviours devoid of control structure' sense used by Preece *et al.* (1994:410), it owes more to the activity theory tradition in application to HCI (Nardi 1996). It connotes actions taken over time, in pursuance of an agent's aims, that are sensitive to the resources available to the agent, including other agencies and artefacts through which those aims are advanced, and the context of action. An understanding of and continuous reactive sensitivity to the roles and activities of other agents in the work system is thus intrinsic to this 'activity' notion. Collaborative work described at this activity level is marked by points of contact and/or handover—actual occurrences of work—but broadens the idea of contact to include potential interactions. Work occurs over varying time frames, including dealings with many different people and serves a variety of purposes. Identifying potential interaction points sets the conditions for reasoning about spontaneous interactions and stands to avoid unnecessary constraint on contact and joint work that may result of making formal and inflexible the apparatus of collaboration.

Where phases of separate and conjoint activity are coupled in this way, requiring mutual consideration of interleaved activities, those activities may be considered as semi-synchronized. Semi-synchronization sets up the idea that co-ordination in collaborative work is not an absolute matter, but joint activity moves along a dimension of coupling, including a level of potential involvement subject to initiation on the basis of cue monitoring. Semi-synchronized activities describe an awareness of potential collaboration as well as current or ongoing collaboration. It further allows for collaborators' other activities outside of the immediately conjoint effort to be included in the performance of the joint activity. Since collaboration is described on a continuum, semi-synchronized activity lies on a dimension of mutual involvement between fully synchronized and unsynchronized extremes.

The claims made concerning the characteristics of collaborative work are articulated with reference to case studies of teleconferenced medical consultations, as an example of technologically mediated collaborative work. These are introduced in §2. In §3 several descriptive requirements of analysis for collaborative work are presented. Section 4 describes the Comms Usage Diagram, a notation incorporating these requirements as a 'framing representation' for further, detailed analysis. It demonstrates how these general co-ordination requirements might be used to describe issues for work in particular collaborative contexts and thereby how to guide appropriate technological design.

2. The problem: a case study

The representation to be described in §4 was developed over a series of three field studies of medical teleconsultation. The nature of this consultation, the personnel involved and the technology available to them are described in this section as an illustration of the kind of collaborative work this paper is concerned to relate to task analysis.

Existing customers of a British telecommunications company's video communication products were approached and asked to take part in the field studies (BT

Laboratories, Advanced Media Unit, Martlesham Heath, Ipswich, UK). Three sites were visited. At each site the work concerned was 'telemedical consultation', that is, the provision of advice by a consultant at some distance from the advice seeker. At each of the sites concerned, there was always a medic on the spot with a patient contacting a relatively senior medic at some other location by videophone (Videophone is used generically here to refer to any telecommunications device offering a real-time moving picture of the party with whom the person using that device is communicating). Site A consisted of a link between Peterhead Community Hospital and Aberdeen Royal Infirmary. Site B included links between Belfast Royal Infirmary and Minor Treatment Centres in the Westminster area of London. Site C involved a collection of GP practices and two hospitals in North London, then known as the Teleconferenced Medical Consultations Project (TMC). Although Site C is specifically a research site, all sites employ video links to support the provision of treatment advice by medical specialists for primary care practitioners, with the consent of the patients concerned, in the course of discharging their regular medical responsibilities.

2.1. *Method*

Data were drawn from three sources: video recordings of the equipment position and layout; talk through of equipment use by an experienced user; and, the primary source, through interview. Interviews were conducted using an interview schedule. This schedule was designed to take the informant through a 'typical' consultation, from switching on the equipment right through to completing any necessary records following the consultation. Seventeen informants were interviewed in this way, including consultants, registrars, senior house officers, GPs, practice managers, nurse practitioners, and registered nurses. The majority of these were audiotaped for later review. As Bødker and Grønbaek (1991:316) have observed: 'Human beings possess skills that are not easily articulated or imagined without actually being in a situation of acting'. Whenever possible, informants were interviewed at the place where the relevant equipment was installed, talking through their experiences and indicating specific points with reference to that equipment. Video recordings were made of these demonstrations. Additionally, a video recording of a consultation was made by one of the authors and a further collection of four videotaped consultations was made available to the authors.

Following each visit, a report was prepared including the following information: the location of the videophone and the parties involved; the purpose of the videophone's installation; the benefits and demerits of working with the videophone; the technological benefits and demerits of the videophone as implemented; a graphical account of a typical consultation session; and an overall higher-level summary of the positive and negative points concerning work practice with the videophone. Copies of these reports were made available to the informants for approval and/or amendment. It is worth noting that several of these did take the trouble to comment, indicating that the reports themselves were accessible and, given that the comments were all minor amendments, that the information included represented a fair summary of the work as carried out in the opinion of the informants. The 'graphical account' was developed iteratively over the three studies and forms the basis of the CUD notation described in §4.

2.2. The work context

In each case, telemedical consultations occurred when it was not clear either that the patient could be dealt with locally, or should be referred on. Thus, these patients in some sense represented 'grey areas' of the primary care practitioners' expertise.

2.2.1. Setting up the consultation:

For sites A and B the consultants were in an Accident and Emergency department and had agreed to be permanently 'on call'. For site C, the telemedical consultations were arranged by appointment only. The patient would simply come to their local practice rather than travel to see the consultant. The consultant receives a referral letter in the conventional manner and thus has some idea of the nature of the case prior to the consultation. One of the on-call sites (A) had a teleradiography system in place, such that the consultant could view a patient's X-ray picture for a few moments prior to the consultation.

2.2.2. Medical expertise and independence:

Site B was a walk-in clinic staffed by nurse practitioners rather than GPs. Here, the presence of the video link was one of the major factors supporting the existence of this relatively new model of primary care. It is perhaps material to note that the skills gap, for the purposes of consultation, between nurse practitioners and consultants is large in comparison to the differential between a GP and a consultant. Furthermore, whereas GPs have a well-established model of care and considerable independence in treatment, nurses are used to working in a support role.

2.3. The technological context

All sites used videophones running over ISDN-2 lines (128 kbits/s). In each case, two kinds of audio connection were available through the videophone: using a conventional telephone handset or a built-in loudspeaker and microphone combination. Video was primarily provided by a CCD colour camera mounted over a display screen. For sites A and B, these were built into a device with a 10" colour screen (BT VC7000). Site C used a PC fitted with a special extension card for video conferencing (BT VC8000) and so displayed images on the PCs 16" monitor, in a resizeable window. Conventionally, the window was always maximized. Again, in each case, it was possible to use an external video feed instead of the main camera. Sites B and C were equipped with adapted camcorders (Sharp Viewcam) for this purpose. When viewing a locally sourced video, the screen displayed the live feed from the camera (i.e. not encoded and decoded). Site A additionally had a specially adapted image scanner for digitizing X-rays, attached to a dedicated PC. The PC was used to display the digitized X-rays and also to transmit them to the remote site (A&E department) via the videophone's ISDN-2 link.

2.4. Summary of collaborative application

This, then, is the kind of technological support for work that the authors are seeking to reason about with their analysis of collaborative activity. A remote expert advises someone about some visible problem. Here the visible problem is a patient but we have in mind parallel applications where the remote and local participants are engineers and the visible problem is some equipment (Kraut et al. 1996). Such applications of video technology should also be amenable to this approach. The reason for carrying out the analysis is to assess the suitability of the videophones used at present to mediate this collaborative work with an eye to making recommendations for new configurations of video technology that may be better suited to the task.

3. Semi-synchronization in telemedical consultation

Describing work for any purpose requires that the approach settled upon is sensitive to characteristics of the work relevant to the purpose for which the description is undertaken. Where the purpose is the introduction of new technologies to some work system, this requires that something is known about available technologies, some ideas about what constitutes the work in question and how work and the available technologies might fit together. This paper is concerned with the design of technology to support work that is semi-synchronized. As described in §1.6, semi-synchronized work casts collaboration as a process of moving between various degrees of mutual involvement in some activity. As was also discussed, the process of coming into contact is understood as occurring through an awareness of the status of the colleagues as they carry out their own activities taking place at any given juncture. This in turn depends on concepts of 'time' and 'space' within which collaboration occurs. By this, time is meant in the sense that to be aware of someone's status one must be working within the same time frame as that status applies. For example, to see them engaging in some activity relevant to one's own interests, or else to see evidence in the form of a document left open on a colleague's desk. Space is meant as the combination of physical constraints on gathering status information, including both the actual environment in which the participants are working and the technologies that mediate the availability of the information. Several aspects of this awareness can be illustrated by an example drawn from a recording of a consultation at site C. It includes several parties taking part in a number of activities under the broad heading of a consultation, and interacting in various ways over its course. The interactions between people, and the progress of their activities, are crucially constrained by their contemporaneity, and the physical and technological factors governing the level of contact possible with one another.

An early teenage female patient had been suffering from dizzy spells and her general practitioner (GP) had booked a telemedical consultation with a paediatric specialist. The girl arrived at the GP's surgery with her mother and, at the appointed time, the GP placed a videophone call to a local hospital. So the GP, patient and her mother were located in the GP's surgery. The consultant was at the telemedical workstation at the hospital with the technicians responsible for the video conferencing equipment. Technicians received the GP's videophone call and handed over to the relevant consultant. The session began with introductions of the various parties involved and a résumé of the problem. Both GP and consultant were using the videophones' standard issue handsets for speaking and listening to one another.

After her initial briefing, the GP passed the handset to the patient who was sitting in front of a video camera fixed over a PC monitor, upon which was displayed a video image of the consultant. The consultant set about asking the patient more specific questions about the nature of her problem. He asked her a series of questions about the time of day and regularity of her attacks of dizziness, and about her lifestyle.

Section 4 will make a distinction between primary and peripheral parties in interaction that is illustrated here. In terms of a work-oriented task analysis, one might think of the part of the consultation scenario described above as having been

drawn from a 'consultant takes history from patient' task specification. This
identifies the primary participants in this phase of the collaborative activity, namely
the consultant and patient, and what they are doing. A communications-oriented
task analysis, on the other hand, needs also to be sensitive to the other people present
and the awareness of all these parties to one another. The importance of this
sensitivity was illustrated in the scenario when the GP switched on 'hands free' audio
in the surgery. Up to this point the GP and the patient's mother were having their
own conversation, aside from the interaction between patient and consultant. They
could see the consultant but could not hear what he was saying since the use of
telephone handsets had personalized the consultant's speech for the patient. With the
GP's intervention, suddenly everyone in the presence of the videolink's loudspeaker
could hear the consultant. Recognizing the change in audience, the consultant made
an effort to bring the GP and patient's mother back into the conversation by
summarizing his line of questioning:

> I'm sorry I I was just going over (.) with **[the patient]** (.) some of the um (.) ya know just her
> life (.) a little little bit about her school life and then I was =
> > *[Patient smiles*
> > *[Mother moves so that her full face is in frame*
> just moving on to her (.) eating (0.5) and she tells me that she eats fairly well

At this point, the patient's mother was visible to the consultant as she narrowed
her mouth and shook her head vehemently. She had been looking intently at the
floor but the consultant, having noticed this action, picked her up on it by saying
simply: 'Mother shakes her head in the background'. The patient's mother then went
on to contradict her daughter's assertion by saying 'she doesn't she doesn't eat a lot':
This episode shows how a nominal collaboration can consist of people who are
involved to varying degrees in furthering the work and how someone 'at the edge' so
to speak can have a major effect on the course of the work. It also demonstrates how
the resources available to support the collaboration, in this case the mutuality of the
visual and auditory connections, exert a powerful effect on the ability of those who
might be able to contribute. Had the GP not switched to hands-free audio, the
consultant would have had no contradictory evidence about the patient's eating
behaviour.

The consultant went on to ask the patient about some other matters, including
her home life. When people talk, they design their speech for one another in subtle
and automatic ways. This is known as recipient design (Clark and Schaefer 1989).
There are very many words and phrases that might be used at any given moment to
express what a speaker has in mind. These loosely comprise a set that are more or
less adequate for the purposes of the speaker and the form chosen is likely to be
dependent on a variety of considerations, including a sensitivity to such matters as
the formality of contact and not least the likelihood of the audience to understand
the chosen form of words. These matters are particularly important in medical
consultation, as specialists routinely deal with the lay public. Speakers are able to
revise their selection criteria 'on the fly' as information about the effectiveness of
their chosen form of expression is assimilated. Additionally, knowing who can see or
hear you is another part of this awareness. This kind of awareness often will be a
problem when telecommunications equipment is used. As the audio was not reverted
to handset mode, all present in the GP's surgery could hear and contribute to the

dialogue. The consultant chose on the face of it a cumbersome expression to ask the patient about her relationship with her mother: 'and things at home are no different from most of your friends?' This reinforces the matter of the presence and mutual accessibility of all parties, but from the opposite perspective that it can hinder as well as help a particular party to carry out their work. A reading of the consultant's form of questioning is that he wanted to determine whether or not abuse was in question. A difficult matter to approach in a one-to-one situation was compounded here by the audio link having been switched on so that the mother could hear this enquiry about a very delicate matter concerning herself.

One of the most important findings of the field study was the perceived value of having parties who may not be directly involved in the current activity overhear what is going on. These peripheral participants may be: relatives, as above; a nurse or a technician who has to take some later action dependent on the outcome of the current activity, or even a major player such as the GP, depending on the unfolding dynamics of actual system use. To take another example, a patient may be a peripheral participant when a GP and a consultant are discussing treatment. If the patient can overhear this discussion, and the consultant and GP are aware of this, then the patient draws confidence and the later activity when the GP explains to the patient what is required is much facilitated. The importance of overhearing, as a matter of audio accessibility and monitoring, is perhaps best illustrated by the fact that all the sites visited used hands-free audio most of the time, this despite the very poor sound quality that resulted.

A final point to be made concerns the deployment and role of particular resources over the course of a period of collaboration. At site C the standard procedure was for the GP to send a letter to book a consultation so that the consultant could appraise the relevance of the case for his/her own area of expertise. Initially then, the letter is a different-time–different-place communication. However it was found that later on it additionally serves as a co-ordinating device for scene setting between the GP and the consultant, the GP having retained a copy. At this point it is a same-time–different-place resource type with two tokens. Finally, the patient may occasionally have access to the letter by looking over the GP's shoulder. That makes it a same-time–same-place resource. In each case, the communications resource is nominally the same and yet fulfils a different function.

To summarize, for our purpose of reasoning about technology for collaboration, an analysis must allow one to take account of the following minimal set of considerations:

(1) what are the primary activities that constitute the work in question?;
(2) which of these primary activities are interactions between agents (distinguished from interactions with equipment)?;
(3) who else may participate (i.e. who has access to the ongoing work)?;
(4) the contemporaneity of agents' activities (from which the potential for opportunistic interaction might be determined);
(5) the space where the activities are taking place (setting constraints on the kind of interaction possible, e.g. visibility with relative position of people and equipment, and audibility against ambient noise levels); and
(6) how accessibility to primary activities is made available, through the resources that provide relevant information about them and the resources that broker interactions between the primary agents once initiated (e.g.

appointment booking system setting up a later meeting, or an audio-video link for synchronous collaboration).

The Comms Usage Diagram (CUD) described in the next section is a representation designed to frame such an analysis. Its value is illustrated by identifying perceived benefits and demerits of the communications technologies used in the field studies.

4.   The Comms Usage Diagram (CUD)

A common abbreviation for 'communications technologies' in engineering is 'Comms' and the graphical representation proposed will be referred to as a 'Comms Usage Diagram' or CUD. The purpose and general mechanism of the CUD will first be discussed. Subsequently, its representational details are described. To clarify the specifics of each issue raised, reference is made to sections of CUDs generated from each of the field studies described in §2.

4.1. *The function of Comms Usage Diagrams*

A CUD provides a high-level description of work on the basis of informants' accounts through a representation of the aggregation and approximate sequence of activities involved in and between the places where they occur, together with the facilities enabling this collaboration. It is thus a kind of scenario tool, rooted firmly in the data collected about the collaborative work in question. The graphical scheme settled on represents the manner in which the work is carried out using the same descriptive terms adopted by the informants: the people who carry out the work. It should be emphasized that as the CUD is an exploratory tool, it is conceived as a notation for framing a consideration of collaborative activity against the available resources for interagent action. CUDs seek to identify where and how communications occur between collaborating parties in some existing work context. However, the 'Comms' term should be read as a descriptor for all interpersonal communications or signalling that subserve the collaboration under consideration and thereby indicating technological possibilities, as well as those communications technologies currently serving their collaboration. An important function of the notation is that is represents the activities of the agents concerned at each geographical location in parallel with one another. This operationalizes a broad conception of collaboration as a continuous process involving to a greater or lesser extent all those present.

As described in §2.1, the informants were to be presented with a summary of the findings. For this reason, an additional function of the notation was that the descriptive content of these summaries be accessible to a non-technical audience. There is widespread agreement in the HCI literature that successful system design is highly dependent on a close consideration of the target population, the 'users', and their active participation in the design process (Good *et al.* 1984, Hewett 1986, Boehm 1988). Any descriptive strategy should at least lend itself to easy explanation to users.

A final point to make is to reiterate that the CUD serves as the focal point of a work study of some kind, where data from a number of sources may be brought together and synthesized into a common format. Wherever some communicative benefit is identified in the collaborative practice examined, it is with reference to the data gathered rather than having been included purely on a 'what if' basis.

4.2. *Overview of a CUD*

Figure 1 represents a part of the telemedical consultation carried out at site A. There are two locations, the Peterhead treatment room and the Aberdeen Royal Infirmary teleradiology workstation in the Accident and Emergency department. The first two columns specify the activities going on at each of these locations and the people who are present. The next column specifies the 'comms resources' used to mediate these activities. Site A is equipped with a videophone and also a teleradiography workstation at both locations. The latter comprised a PC with an image database and viewing facility at the specialist centre and an X-ray scanner/digitizer with similar PC at the primary care centre.

The CUD can be read from the top to the bottom to expose the approximate order of events thereby building up a picture of what actually happens during the collaborative work. The first activity given in the CUD is 'discusses X-ray', carried out by a GP at the treatment room of a primary care centre, with a consultant at the co-operating hospital. If the consultant has trouble in making out some detail on the X-ray there may be a subsequent activity represented as 'Consultant requests better X-ray image'. In this case the nurse at the Peterhead primary care centre re-scans the original X-ray and transmits it to Aberdeen.

The CUD can also be read left to right, to indicate co-ordination of activities going on at the various locations and to see how the technologies deployed interact with those activities. As indicated above, column 3 in figure 1 specifies the technologies used to mediate the activities specified in columns 1 and 2, that is,

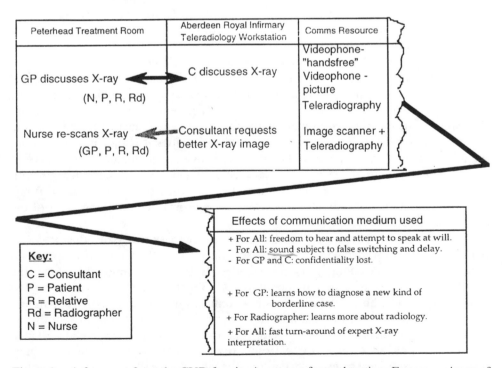

Figure 1.   A fragment from the CUD for site A, see text for explanation. For convenience of printing the diagram, single rows of the CUD are split as indicated by the ragged right-hand edge of the upper segment and left-hand edge of the lower segment.

sound, video images and copies of the transmitted X-ray. Finally, the right-most column of figure 1 allows summary remarks about the communicative value of the technology employed in its support of the work as carried out. Here, it is observed that the audio link employed offers as a benefit that all those present at Peterhead can overhear the discussion between GP and consultant. So for example, the nurse can overhear the discussion of the GP and consultant in the first activity 'discusses X-ray' and so may be able to anticipate the request in the second activity 'requests better X-ray image'. On the minus side it is noted that this overhearing may also lead to occasional confidentiality problems. In addition, there may be problems with echo cancelling that lead to very poor sound quality. Some of the points made in the right-most column are more general. One of the clear benefits of this system is that the GP learns to diagnose X-rays in these borderline cases.

Having provided a broad outline of the CUD the next sections discuss various issues of representation in more detail. They are the representation of time, participation, technological mediation and the effects of that mediation.

4.3. *Representing time in the CUD*

Order is not strongly represented in the CUD. The idea of 'task phase' is used to describe how work is ordered over time, where each task phase comprises collections of those activities taking place within the period. The ordering of activities within each phase is subject to a number of forces and the CUD makes no attempt to prescribe it. The lack of prescribed sequence to the activities in a task phase reflects on the kind of flexibility to be observed in most collaborative encounters. People take advantage of appropriate opportunities to fulfil their objectives as and when they arise. A collaborative consultation will routinely involve revisiting various activities as an understanding of the patient's problem is progressively refined. Also, several relevant activities may occur simultaneously.

It would be possible to represent time explicitly as a continuous variable or as sequential dependencies, as in the 'plan' part of an hierarchical task analysis. However, this would make the representation very much more complex and runs the risk of enforcing order where there is no requirement for strict sequential dependency. As the authors' primary purpose is to associate the merits and demerits of communications technologies with particular activities it is difficult to see what would be gained by doing this. The cost of providing an unambiguous explicit coding of the sequential dependencies between activities would be considerable both in terms of the analyst's time and in terms of overloading the representation. When time becomes an issue it can be included in the right-most column, the effect of the comms resources used.

Consider figure 2 where two task phases are represented. Note that they are marked out by an horizontal line across the full width of the CUD. The first phase may be thought of as 'contact' and the second as 'history and examination'. Since the 'contact' phase is closer to the top of the CUD, it is taken to occur prior to 'examination'. The horizontal rule does not connote an absence of activity spanning phases. In this example, the change from one phase into another is an action of the nurse practitioner (NP), changing the video ons link to show a view from a roving camera to the remote consultant standard camera, fixed above the videophone's screen. The descriptive rises from the accounts of the work given by our informants and a -base that suggests an ordering of the aggregation of tasks. So 'contact' naturally precedes 'history and examination' but, within 'history and examination', there is nothing to be gained by insisting that the joint activity 'discusses case' precedes, or indeed is completed independently of, 'moves camera'.

Order is implied through the content of the activities and, where multi-agent activities have a strong sense of initiation or causation, through the use of directional arrows. In figure 2, the 'contact' phase includes placing the nurse practitioner initiating a call to the hospital from the Minor Treatment Centre, and then a passerby responding by fetching a relevant consultant. Meanwhile, the patient and (potentially) relative wait for the co-operating consultant to arrive before the joint activity greeting and introduction. The 'history and examination' phase consists of discussion and (where appropriate) viewing the patient. So ordering of these events is given by the single-headed arrows where some initiation is required, otherwise it is deemed to be unnecessary (questioning and examination may very well alternate) or else is implied by the content (the roving camera must be switched on before the consultant can direct the nurse practitioner to provide the required image). The CUD represents a collection of activities within a task phase as essentially

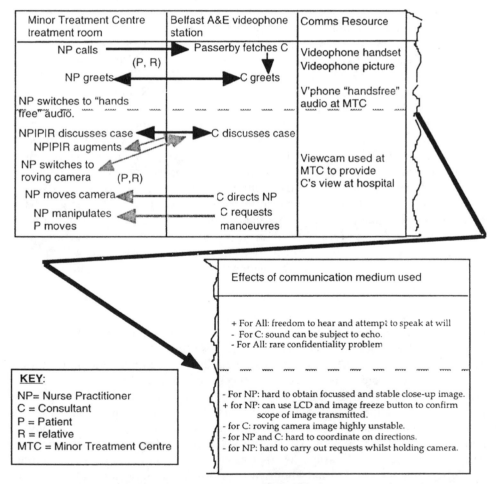

Figure 2.   A CUD fragment from site B, two task phases each containing several activities.

simultaneous streams of activity at each site that couple and decouple through the co-ordination between those involved.

4.4. *Representing participation in the CUD*
4.4.1. *Agents, activity and location*: An appropriate representation of work requires some principle for deciding upon the inclusion of relevant parties and activities. The specifics of relevance are complex, and are interpretable at many levels, from the notion of any conceivable impact on the work of interacting communities through to inclusion in a system specification. Here, the CUD takes presence as the only criterion for relevance to the work, where presence may be in terms of physical co-location at a given time or projected through media. The entities represented in the CUD are in every case a combination of one or more *agents* and *locations*. An agent is potentially any body to which it is sensible to attribute concerted and purposeful action. The most straightforward kind of agent is an individual, the GP, the nurse and so on. In some instances it might be sensible to consider an organizational unit or department as an agent, as it impinges on a place from some other location. For site C, the consultations were organized on an appointments basis. The office dealing with appointments could be used as an intermediary in real-time co-ordination of the teleconsultation in case of equipment failure. In this case the office was treated as an agent.

Figure 2 shows the nurse practitioner (NP) contacting the hospital from her workplace. Her relevance to the consultation is obvious. Also, in the treatment room, although not taking part in the call moment, there will be a patient (P) and possibly also a relative of the patient (R) represented as 'P, R' in parentheses next to that activity. At that moment, their relevance to the consultation is less immediate. The idea of level of collaborative involvement in interaction was discussed in §3. Involvement in an interaction is usually seen as fully reciprocal, where the parties are directly and explicitly involved in the initiation or carrying forward of the exchange. This is referred to as primary participation. Thus, the nurse practitioner in figure 2 is the primary participant in the initial contact with the hospital. An alternative and perhaps more pervasive involvement in interaction occurs indirectly, as peripheral participation. Peripheral parties can influence the interaction between the primary parties, or else derive from their indirect involvement in one interaction some influence on their own subsequent activity. For example, the presence of the patient and relative may very well affect the manner in which the nurse practitioner deals with the hospital.

In the CUD, association with activity is used to distinguish between primary and peripheral participants. Primary participants are named as taking part in the activity while peripheral participants are simply represented as being present during the task phase in which the activity occurs. That is they are in a shared space contemporaneously with the primary participants of activity and may become synchronized or increase their degree of synchronization with it. To take another example from figure 2, in the activity 'discusses case', NP and C are the primary participants as they are mentioned with the activity. P and R are peripheral to this activity, although the fact that they can hear and see what is going on may be important. Specifying an activity with primary and peripheral participants helps to describe how each of the various agents concerned stand in relation to one another. It allows something to be said about each person's readiness and availability for contact at any given moment. It also describes some of the constraints on the ability of any agent to recruit others in pursuit of their aims. By taking care always to represent those present at a location, the CUD at least makes possible reasoning about peripheral as well as primary interaction. All agents appearing in a CUD are automatically guaranteed relevance by their very presence and active identity.

4.4.2. *Control and initiation*: The CUD represents primary interaction through the use of arrows. The default characterization is given by a double-headed arrow linking more than one agent, their part in the interaction given by the name of the activity with their identifier (GP, P, N, etc.) The activities of agents linked by such arrows are to be read as properly synchronized in that all linked parties contribute to the execution of the task. The linking of two activities in this way implies that they are deliberately co-ordinated and the 'space' in which they occur. Examples of such fully synchronized tasks are to be found in figure 3. This represents 'introductions' and 'history taking' task phases at site C. It shows a GP, patient and (possibly also) a relative introducing themselves to the consultant at the remote hospital. They are each directly involved in this encounter. During 'history taking', the GP or the patient may interact with the consultant.

In contrast, single-headed arrows include a notion of precedence, and hence explicitly represent a sense of initiation or control. In other words, the activity occurring at the source of the arrow not only was put in motion by its associated agent, but also brought about the activity given with the agent at its destination. They indicate direction rather than collaboration. Figure 2 shows that the

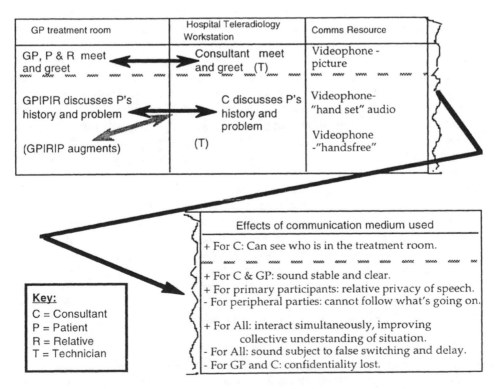

Figure 3. 'Contact' fragment of site C Comms Usage Diagram.

'examination' phase at site B involved the consultant giving directions to the nurse practitioner, firstly to manoeuvre a roving camera into a position to provide a view of the patient required by the consultant, and then to carry out some manipulations of the patient. In contrast, all the interactions represented in figure 3 are shown as fully collaborative in nature.

Control in a wider sense is a fundamental issue in medicine, as it is vitally important that the physician responsible for a patient be clearly identified. The CUD does not represent control at this level. Similarly, single-headed arrows are not to be read as evidence of authority.

4.4.3. *Additional activity and peripheral involvement*:   Although it was a fairly common request at site A, the consultant will not always ask for a clearer X-ray. This is signified in figure 1, by using a grey arrow rather than a black arrow. Similarly, the use of the roving camera at site B (figure 2) was not inevitable and thus the activity 'switches to roving camera' is linked to the context for its occurrence (namely; the joint activity of 'discusses case') by a grey arrow. Consequently co-ordination of its use ('C directs NP' and 'C requests manoeuvres') is contingent on the roving camera being called into action.

In figures 2 and 3 there are grey arrows associated with the 'discusses' activities. (Site B 'discusses' activity is represented as 'case' as it was a walk-in clinic, and so there was neither a history of contact with the primary medic nor medical records for the patient available in this context. Site C 'discusses' activity is represented as 'history and problem', since by appointment these resources could be drawn upon.) These signify interactions that are clear augmentations to the idealized 'task line'. Peripheral involvement is given implicitly in the CUD through representation of presence and accessibility. As previously discussed, collaborative involvement is seen as a dimension, not as an absolute matter. In figure 3, a technician is shown in the company of the consultant at the hospital. In some circumstances, it might be that this presence inhibits the patient from speaking freely but this is to some extent unpredictable. Some kinds of involvement are peripheral but well within the remit of the collaboration. When the consultant interacts with the patient, the GP may add information to their conversation as appropriate, without necessarily taking the floor. During the consultation reported in §3, the consultant suggested that the patient's problem may be due to a build-up of ear wax. The GP, having heard this suggestion, immediately produced an appropriate instrument and checked the patient's ears, without making any attempt to take the floor until she had determined that there was no wax problem. While the consultant is interacting with the GP the patient or relative may also chip in, adding to the primary activity of 'GP and C discusses P's history and problem'. These then are anticipated possible lines of primary participation, originating from peripheral involvement. The peripherality is given by parentheses and potential primary interaction by grey rather than black arrow links. They say something rather stronger about the nature of peripheral participation than the bracketed tokens of presence in figures 1 and 2, in that they anticipate the kind of activity in which these peripheral parties are likely to be engaged.

4.5. *Representing technological mediation in the CUD*
The CUD allows for reasoning about contact and conjoint activity through the awareness of participating agents in a collaborative workspace. It specifies which media are used to support which tasks. At each point where activity is represented as

occurring between different locations, the media through which this activity is carried out are specified in the column headed 'Comms resource'. These are given in the first place at the same horizontal level as the activity to which they refer. Activities given further down the CUD are taken as inheriting from the introduction of these technologies in the 'Comms resource' column. Any change to the configuration is then noted next to the activity or activities at which the change was made.

Where agents are at the same location, i.e. co-present, it can generally be assumed that they can see and hear each other, although this is not inevitable. It may very well be in some circumstances, for example in a factory, that co-presence is not enough to guarantee accessibility. People may not be able to hear one another clearly because of noises. In the three sites studied there were no such difficulties, although it was not the case that co-presence guaranteed being able to see one another. An additional representation was developed in order to reason about the potential awareness of one participant to another; this is the participant-percept matrix. An example of a participant-percept matrix is given in table 1.

In table 1, the horizontal rule between the row for the audibility of the nurse and the row for the audibility of the consultant signifies that the percepts refer to sources from different locations. The GP, patient and nurse are all co-present and can all hear each other well (C+). They are using the handsets rather than 'hands-free' audio, therefore only the GP can hear the consultants voice (E+). The way that they are seated enables the GP and patient to see each other's faces (C+) and the consultant's face (E+). The nurse may not have access to these percepts, depending on what he or she is doing at the time (C? and E?, respectively).

The percepts chosen for a participant-percept matrix will depend on the tasks being mediated. The faces and voices of all the participants will normally be included: the voice because it carries the conversation and the face because it allows

Table 1. Participant-percept matrix from site A. Configuration c1—Consultant and GP using handsets, GP in front of camera.

| Percept | Participant | | | |
	GP	Patient (P)	Nurse (N)	Consultant (C)
Hear GP voice	C+	C+	C+	E+
Hear P voice	C+	C+	C+	
Hear N voice	C+	C+	C+	
Hear C voice	E+			C+
See GP face		C+	C?	E+
See P face	C+		C?	E+
See N face		C?		E?
See C face	E+	E+	E+	
See X-ray	E+	E?	E?	E+
See P's problem	C+	C+	C?	

Key: C+, copresent, can hear or see; C?, copresent, can sometimes hear or sometimes see. E+, Electronic, can hear or see; E?, Electronic, can sometimes hear or sometimes see. Empty cells indicate that the percept is not available to the participant.

one to identify the other person and to determine their focus of attention, for example. In table 1, two additional percepts are included, 'See X-ray' and 'See P's problem'. See X-ray refers to the accessibility of the teleradiology workstation. See P's problem refers to occasions when the patient would be moved in front of the video telephone in order to show the consultant a wound or whatever. With the particular configuration described in table 1, with the GP in front of the camera, this percept was not available to the consultant.

For many interactions it is unnecessary to generate a participant-percept matrix as the information it conveys is implicit in the content of the third column describing the comms resource used. When there are several possibilities, such as here (e.g. handset versus hands-free audio, and GP in front of camera versus patient in front of camera) then it may be valuable to spell out what percepts are accessible in a participant-percept matrix for each of these possible configurations. References to these configurations are then added to the CUD in the comms resources column.

The authors have concentrated mainly on the use of audio and video technologies but the CUD can also be used to describe communication using other technologies such as text. It is easy to think of these negatively, as simply 'narrowing the bandwidth'. For example, one might at first only imagine that using a letter means that synchronous clarification of the message it contains is precluded. However, it can also bring other changes to the contact resources by enriching the available pool. Letters are persistent. They may be reviewed and detailed information may be closely scrutinized. Following on the letter example, the CUD may also be used to explicitly represent media that support communications within workspaces over time, where co-presence is not to be had. For example, site C's booking system included the continuance of the convention to send a referral letter to the consultant. As discussed in §3, this letter could be used by the GP herself as a reminder of the salient points she wished to make, as well as serving as a shared resource to co-ordinate and give structure to a discussion.

4.6. *Representing the effects of communications technology*

The final part of the CUD is to include notes on the particular advantages and disadvantages of the resources employed for each of the identified activities in the right-most column of the CUD. They record whether or not the point raised is seen as positive or negative, the agents for whom it is good or not so good, and the way in which it operates. As was stated in §4.2, the comments vary in scope. Some are very narrow and relate to specific features of the equipment, such as the note in figure 2 about the use of the LCD panel on a roving camera. Some are much broader and refer more to the social situation that the equipment permits, such as that the patient was said to be much reassured by the presence of the GP in their meeting with the consultant (without teleconsultation, the patient would have seen the consultant on a one-to-one basis).

These notes are the major reason for performing the analysis. The suggestion that some facility provides some advantage or disadvantage will normally originate directly from a comment by an informant or observation of the work. What the CUD adds is the possibility to locate the advantage or disadvantage more precisely within the work and to find other places in the work that could be affected similarly. In the case of advantages one can then see what steps could be taken to configure the technology and the work so that they apply as widely as possible. In the case of disadvantages one can reason about alternative ways of organizing the work and configuring the technology to minimize their impact. The way that the CUD is used is discussed further in the next section.

5. Discussion

The earlier sections of this paper characterized a particular kind of collaborative work described here as semi-synchronized activity. This characterization arose from a consideration of the task analysis and CSCW literatures. The stimulus for this synthesis was the problem of analysing the influences of the communication technologies used in telemedical consultation.

The CUD is a representational framework for reasoning about technological support in semi-synchronous activity. The CUDs generated in the three field studies were invaluable in integrating the information obtained from different informants and as well as from direct observation. They made it possible to locate the comments to particular activities and so to examine the completeness of the resulting account. The account was compact, two sides of A4 for two sites and one for the other, yet specified the activities undertaken in some detail.

The CUDs have also demonstrable value in making recommendations about the communications technologies used (Watts and Monk 1997). For example, many of the effects identified in the last column concerned sound. Considerable advantages were seen for overhearing, implying a need for multi-party sound, but at the same time the poor quality of current facilities was seen to impede work. A strong recommendation can thus be made for more engineering effort to be put into echo cancellation and other measures to make multi-party sound more effective. Another recommendation coming directly from the CUDs was that the consultant in the hospital should have access to a wide-angle view of the whole treatment room where the patient and GP are located. This is so that the consultant is fully aware of all the people that may overhear or who have the potential to join in the conversation. This requirement presents several interesting human factors and engineering problems concerning how one transmits and presents multiple views. Again, by locating the effects to particular activities it is possible to reason about the alternative solutions to these problems.

The paper started with a discussion of the pragmatic goals of task analysis. These are to abstract out the characteristics of some work situations necessary for effective intervention. The practical problem of devising better communications facilities for collaborative activity at a distance was shown to pose a challenge to existing approaches and the CUD can be seen as a step towards answering this challenge. In the CSCW literature reviewed in §1.4 many authors have criticized the quest for abstraction in the description of work. The danger is that the act of generating the abstraction restricts the view of the analyst. The aim here has been instead to enhance the analyst's view. By de-emphasizing sequential constraints and emphasizing the potential for participation, the CUD identifies opportunities for collaboration and does not prescribe a particular way of working. In that sense the notion of semi-synchronized activity serves to bring together some of the aims of both the task analysis and CSCW communities.

The CUD, as described in this context, has been employed to identify some general connectivity and awareness issues for a particular form of telemedical consultation. It has further been emphasized that the functions of an adequate technology to support this work should *at least* include those described above. Mindful of the sort of concern that led Bowers and Benford (1991:343) to comment that 'all representations have limits . . . [they] are always incomplete and their pursuit can sometimes lead to ruination', it deliberately does not make these recommendations as a blanket prescription for this work. It has gone to some lengths to describe how work and tools of work interact throughout the course of collaboration, such that advantages at one stage are less of an issue, or even a demerit at others. Furthermore, the model of collaboration beneath the CUD is not intended to support a complete account of collaborative work. It is described as a framing representation, rather than as a comprehensive description given by an exhaustive analysis.

Any representation is of limited value without well specified procedures for producing and using it. The next step in the authors' research is to make explicit the procedures used in the field studies described here and to test them in other contexts. One avenue to be explored, with regard to procedures for using a CUD, is that of

Table 2. Guidelines for the potential value of percepts.

Percept	Potential value	Requirements
Hear voice	Necessary for principle participants to share information	Good quality sound
	Overhearing allows a third party to: (i) interject (e.g. patient corrects GP on point of fact); (ii) understand things they also need to know without the need for repetition; (iii) gives confidence	Poorer quality sound acceptable
See face	Necessary to know who is present (important for recipient design)	Slow scan, low definition acceptable
	Useful prompt to memory for previous conversations	Slow scan, low definition acceptable
	Useful to see focus of attention of participant	Slow scan acceptable, medium definition
	Useful to disambiguate poor sound	Full motion, high definition
See own face	Participants want to know what the remote participant can see (e.g. may move out of frame, may be worried of unflattering image being seen)	'Confidence' monitor
See X-ray or P's problem	Image must be of sufficient quality for the job	Required definition will depend on the job. Slow scan acceptable
	Participants sharing visual information need to know what the others can see particularly when that image is poor	'Confidence' monitor
	Language is much more efficient when participants can point	'Virtual' pointer

guidelines concerning the potential value of different percepts for given work contexts. These could be applied to participant-percept matrices in conjunction with the rest of the CUD. Table 2 lists some possible guidelines that could be used in this way. The procedure followed would involve considering the relevance of each 'potential value' in the table to a particular activity listed in the CUD. The participant-percept matrix for each possible communication configuration that could obtain with that activity would then be examined with regard to the availability of the percept. This in turn would lead to recommendations about which configuration should be used when and to the specification of new, more advantageous, configurations.

Acknowledgements

This work was supported by the ESRC Cognitive Engineering programme. The authors would like to thank all those who have contributed to this study at the following organizations: BT Laboratories, Royal Free Hospital, Virtual Outreach project, the South Westminster and Parson's Green Minor Treatment Centres of the Riverside NHS Trust, together with Belfast Royal Infirmary, and the Peterhead Community Hospital and Aberdeen Royal Infirmary. The authors would also like to thank the anonymous reviewer whose many helpful suggestions were instrumental in improving the clarity and focus of this paper.

References

BAECKER, R. M. 1993, *Readings in Groupware and Computer Supported Cooperative Work: Assisting Human-Human Collaboration* (San Mateo, CA: Morgan-Kaufmann).

BANNON, L. 1995, Editorial, commentaries and a response in the Suchman-Winograd debate, *Computer Supported Cooperative Work*, **3**, 29.

BANNON, L. J. and SCHMIDT, K. 1991, CSCW: four characters in search of a context, in J. M. Bowers and S. D. Benford (eds), *Studies in Computer Supported Cooperative Work: Theory, Practice and Design* (Amsterdam: North-Holland), 3–17.

BØDKER, S. and GRØNBAEK, K. 1991, Cooperative prototyping studies, in J. M. Bowers and S. D. Benford (eds), *Studies in Computer Supported Cooperative Work: Theory, Practice and Design* (Amsterdam: North-Holland), 315–332.

BOEHM, B. W. 1988, A spiral model of software development and enhancement, *IEEE Computer*, **21**, 62–72.

BOWERS, J. M. and BENFORD, S. D. 1991, *Studies in Computer Supported Cooperative Work: Theory, Practice and Design* (Amsterdam: North-Holland).

CARROLL, J. M., KELLOGG, W. A. and ROSSON, M. 1991, The task-artifact cycle, in J. M. Carroll (eds), *Designing Interaction: Psychology At the Human-Computer Interface* (New York: Cambridge University Press).

CLARK, H. H. and SCHAEFER, E. R. 1989, Contributing to discourse, *Cognitive Science*, **13**, 259–294.

DRAPER, S. W. 1993, The notion of task, in S. Ashlund, K. Mullet, A. Henderson, E. Hollnagel and T. White (eds), *Proceedings of the INTERCHI'93 Conference on Human Factors in Computing Systems* (Reading, MA: Addison-Wesley), 207–208.

DRURY, C., PARAMORE, B., VAN COTT, H. P., GREY, S. and CORLETT, E. 1987, Task analysis, in G. Salvendy (eds), *Handbook of Human Factors* (New York: John Wiley), 370–401.

FILKES, R. E. 1982, A commitment-based framework for describing cooperative work, *Cognitive Science*, **6**, 331–347.

GOOD, M. D., WHITESIDE, J. A., WIXON, D. R. and JONES, S. J. 1984, Building a user-derived interface, *Communications of the ACM*, **27**, 1032–1043.

GREATBATCH, D., HEATH, C., LUFF, P. and CAMPION, P. 1995, Conversation analysis: human-computer interaction and the general practice consultation, in A. F. Monk and N. Gilbert (eds), *Perspectives on HCI: Diverse Approaches* (London: Academic Press).

HEWETT, T. T. 1986, The role of iterative evaluation in designing systems for usability, *HCI'86 Conference on People and Computers II* (Cambridge: Cambridge University Press), 196–214.

JOHNSON, P. 1992, *Human-Computer Interaction: Psychology, Task Analysis and Software Engineering* (Maidenhead: McGraw-Hill).

KIRWAN, B. and AINSWORTH, L. K. 1992, *The Task Analysis Guide* (London: Taylor Francis).

KRAUT, R. E., MILLER, M. D. and SIEGEL, J. 1996, Collaboration in performance of physical tasks: effects on outcomes and communication, in M. S. Ackerman (ed.), *CSCW96*, Boston, MA (Amsterdam: ACM Press), 57–66.

McDANIEL and BRINK, 1997, Awareness in collaborative systems: A CHI 97 Workshop, *SIGCHI Bulletin*, **29**, 68–71.

MILLER, R. B. 1962, Task description and analysis, in R. M. Gagne (ed.), *Psychological Principles in Systems Development* (New York: Wiley).

NARDI, B. A. 1996, *Context and Consciousness: Activity Theory and Human-Computer Interaction* (Cambridge, MA: MIT Press).

NARDI, B. A. and MILLER, J. R. 1991, Twinkling lights and nested loops: distributed problem solving and spreadsheet development, *International Journal of Man-Machine Studies*, **34**, 161–184.

NARDI, B. A. H. S., KUCHINSKY, A., LEICHNER, R., WHITTAKER, S. and SCLABASSI, R. 1993, Turning away from talking heads: the use of video-as-data in neurosurgery, *InterCHI'93 Conference on Human Factors in Computing Systems*, Amsterdam, The Netherlands (Amsterdam: ACM Press).

PREECE, J., ROGERS, Y., SHARP, H., BENYON, D., HOLLAND, S. and CAREY, T. 1994, *Human-Computer Interaction* (Reading, MA: Addison-Wesley).

ROBINSON, M. 1991, Computer supported co-operative work: cases and concepts, reprinted in R. M. Baecker (ed.) 1993, *Readings in Groupware and Computer Supported Cooperative Work: Assisting Human-Human Collaboration* (San Mateo, CA: Morgan Kaufmann), 29–49.

ROBINSON, M. 1993, Design for unanticipated use, in G.D. Michelis, C. Simone and K. Schmidt (eds), *Third European Conference on Computer Supported Cooperative Work*, Milan, Italy (Dordrecht: Kluwer), 187–202.

ROSENBURG, D. and HUTCHINSON, C. 1994, *Design Issues in CSCW* (London: Springer-Verlag).

SHEPHERD, A. 1995, Task analysis as a framework for examining HCI tasks, in A. Monk and N. Gilbert (eds), *Perspectives on HCI: Diverse Approaches* (London: Academic Press), 145–174.

TAYLOR, F. W. 1911, *Principles of Scientific Management* (New York: Harper).

WAERN, Y. 1989, *Cognitive Aspects of Computer Supported Tasks* Chichester: Wiley).

WATTS, L. A. and MONK, A. F. 1997, Telemedical consultation: task characteristics, in S. Pemberton (ed.), *CHI'97 Conference on Human Factors in Computing Systems*, Atlanta, Georgia (Amsterdam: ACM Press), 534–535.

WATTS, L. A., MONK, A. F. and DALEY-JONES 1996, Interpersonal awareness and synchronization: assessing the value of communication technologies. *International Journal of Human Computer Studies*, **44**, 849–873.

WINOGRAD, T. and FLORES, F. 1986, *Understanding Computers and Cognition: A New Foundation for Design* (Norwood, NJ: Ablex).

CHAPTER 5

Issues of quality and practicability in task analysis: preliminary results from two surveys

Les Ainsworth and Ed Marshall

Synergy Consultants, Royal Oak Barn, Cartmel, Cumbria LA11 6QB, UK

Keywords: Control room; Task analysis.

Surveys of task analysis studies in two different industrial sectors within the UK, defence and nuclear power, have shown that there can be wide variability between them, both in terms of level of reporting and the depth of analysis. Particularly prevalent shortcomings were that data sources and methods were often either not specified or were described in insufficient detail. Thus users of the information could be asked to accept potentially costly design recommendations with little indication of the strength of the evidence, or of the effects of taking no action. Some tentative conclusions are drawn about the use of specific task analysis techniques but, while these are useful pointers, more evidence is required. Therefore, further study of task analysis techniques is needed in order to provide analysts with better guidance regarding the selection and use of task analysis methods and the reporting of task analysis findings.

1. Introduction

In order to ensure that complex systems are designed and operated in a safe and effective manner, it is necessary to take a task-based analytic approach that focuses upon the way that all the interfaces (hardware, software, procedures, training, organization, etc.) support the user to accomplish the required tasks. This necessitates gaining a comprehensive understanding of the behavioural and cognitive requirements of the tasks and then assessing how well the various interfaces contribute towards meeting these requirements. In other words, some form of task analysis (TA) will inevitably be necessary to design or assess the adequacy of the interfaces in any complex system that is operated by humans.

There are many tools and techniques that can be used to gather task-related information and place it into a meaningful context. Each of these has its particular strengths and weaknesses. Therefore, the success, or otherwise, of any TA exercise will to a large extent depend upon how the mix of TA techniques and data sources is selected and applied. Ideally, for any TA project, there should be an overall plan and subsequent close monitoring to ensure that objectives are met. Ainsworth and Pendlebury (1995) provide a good illustration of the importance of such close management.

Each TA exercise must be tailored to answer different questions on widely differing systems and this potential variety of applications means that it can be difficult to select and apply the methods appropriately. Informal discussions with analysts, and with those responsible for commissioning and managing TA exercises, suggest that there is potential to improve the efficiency and effectiveness of the practical application of TA. Furthermore, it would be helpful to have some evidence about the level of variability that can be expected in the quality of TAs. Although TAs can provide the psychological insights that ensure that a system fully supports the users and enables them to undertake their tasks effectively, unfortunately, it is also possible for the outputs to be little more than incomplete task descriptions that give scant consideration to the feasibility of the identified tasks.

In view of this, two surveys were initiated in mid-1996 to examine TAs in the UK defence and nuclear industries. In both surveys the main aim was to examine the ways in which TAs were conducted. Key aspects of interest were the extent of variability between different TA exercises and the suitability of different TA methods.

Some 90 studies were considered in the two surveys and these encompassed a wide range of tasks. Most of them addressed tasks in which human operators carry out complex activities involving high level cognitive skills when interacting with advanced technology in time-critical situations. In the defence area, the aim was generally to develop and assess new systems and to analyse tasks to ensure a close match between human abilities and projected operational demands.

In the nuclear industry about one-third of the TAs were undertaken to aid design decisions and the remainder were validations of existing systems. The analyses aimed to provide systematic evidence that operational tasks, with critical human involvement, whether routine or emergency, could be completed effectively with the provided equipment and within the permitted time. The majority of these tasks were highly proceduralized control room-based tasks, but they also involved some local-to-plant and maintenance operations.

The surveys sought to identify good practice in TA as well as difficulties, in order to develop more systematic guidance for the selection and application of TA methods and particularly to ensure that the findings from TAs can be presented in the most effective manner. In this paper some preliminary findings from these surveys are presented and discussed.

2. Defence industry survey

The first survey was a scoping study undertaken in the defence industry. The key objective was to develop an overview of the way that TAs were executed and reported. This was seen as a precursor to the production of prescriptive guidelines. The survey was based upon a representative sample of TAs undertaken by different organizations for all three armed services.

A list of potential TAs was developed after discussions with ergonomists involved in defence-related TA. For commercial reasons, it was not possible to review all of the identified studies, but 27 reports were obtained, of which 18 provided detailed accounts of TAs. The available reports were reviewed using a pro forma. This pro forma was divided into sections that sought information on various aspects of each study, including its purpose, data sources used, analysis techniques, information presentation, and resources required. Where possible, the analysts and those responsible for commissioning the studies were contacted to elicit additional data and clarification. For six of the studies, where the actual reports could not be obtained, the analysts agreed to take part in semi-structured interviews where they answered questions about their methodology and general approach. The results of the survey are summarized in table 1.

2.1. *Reasons for the TAs*
The reasons for undertaking each TA were determined from the reports. These reasons were then verified by discussions with those responsible for managing or

Table 1. Summarized results from studies in the defence survey.

Reason for study	No. of studies	Methods	Data collection	Presentation
Systems procurement	2	HTA, modelling	Technical expert interviews, informal discussion, scenarios	Decomposition tables, subjective reports
Allocation of function and workload assessment	9	Workload analysis, functional flow analysis, questionnaires, rating scales, modelling	Technical reports, direct observations, expert discussions, subjective reports, taxonomies	Flow charts, timelines, verbal transcripts, workload assessment graphs
Manpower analysis and personnel requirements	7	HTA	Walk-throughs, interviews, expert discussions	HTA diagrams
Interface design	9	HTA, link analysis	Informal discussions, technical documents, questionnaires, rating scales	Decomposition tables, questionnaire results
Operability	5	HTA, cognitive modelling	Walk-throughs, expert discussions	Decomposition tables
Systems modelling	5	Micro-SAINT, cognitive modelling	Simulator trials, scenarios	Timelines, decomposition tables
Training	2	HTA, taxonomies, error analysis	Direct observations, expert discussions, questionnaires	
Human error identification	1	Qualitative assessment only	Expert discussions	

undertaking the studies. Eight broad reasons were identified for embarking upon TA and these are listed in the table. Generally three or more reasons were declared for undertaking each study. It was also clear that in many cases the TAs dealt with a wider range of issues than was envisaged when they were first commissioned. In a minority of these studies, this widening of the basic reasons for the TA was apparently due to poor initial project definition. However, in those TAs, where the initial purpose was clearly defined and yet there was a subsequent broadening of objectives, it was not clear whether the additional derived information was of any value (i.e. did the information relate to the declared issues of prime importance?).

As table 1 shows, two TAs were carried out in connection with systems procurement and as such aimed at assisting in the conceptual development of new systems, for example, ascertaining the level of expert assistance that would be feasible in command/control centres. Nine of the TAs were undertaken to assess workload or to optimize the allocation of functions between humans and machines. Typically these studies sought to measure dynamic change in workload during time-critical mission scenarios.

Seven TAs aimed at establishing manning requirements in terms of crew complement and personnel characteristics. Hierarchical task analysis (HTA) was universally applied in these studies.

Nine of the studies were undertaken to assess the adequacy of software or hardware interface design, for example, three TAs evaluated flight deck design innovation. The interface studies were generally based upon detailed HTAs, followed by a task decomposition of the lowest level task elements.

Five of the studies were undertaken to assess whether proposed systems would meet predefined operability criteria. In these studies the key tasks were adequately identified by a combination of walk-throughs on earlier versions of similar systems and discussions with technical experts (including operators). Most of these TAs started from the standpoint that existing systems were operable, then they compared TAs of existing and proposed systems to identify any differences that might result in higher workloads. Finally, they assessed the impact of any additional workload. A notable exception to this was a study that, very effectively, used a basic cognitive model based upon a series of well-defined task elements to assess operability against quantitative criteria.

The five studies that were undertaken to develop system performance models generally provided models that tended to be very scenario-specific, but were, nevertheless, useful. However, the resources that were necessary to develop these models were relatively high, both when detailed cognitive modelling was adopted and when commercially-available computer packages were used.

Only two studies were aimed at training requirements and just one addressed human error identification.

2.2. *Data sources*

A serious shortcoming with many of the reports was that data sources were often not clearly defined. This meant that it was difficult to decide how much confidence should be placed upon findings and recommendations. Typical problems that were noted included:

(1) completely omitting any references to data sources, or inadequately defining them;
(2) not defining the background or experience levels of personnel who had provided technical or operational information; and
(3) omitting key details, such as whether trials had been in simulators rather than operational systems.

As regards specific data sources, although audio and video recordings proved useful as a back-up source of information, it was clear that it was extremely resource-intensive to use them as a prime data source. It was also surprising to note that both the reports themselves and the discussions with the analysts who could be contacted, indicated that very little use was made of written procedures, although these often provide a rich source of information in TA (Pendlebury and Ainsworth 1992).

Simulators proved to be a much-used source of task data. In many of the studies they were utilized very effectively, especially when the analysts took the effort to plan their data collection so that normal training sessions could be used. However, there were also some clear instances where simulators were considered to be an extravagant use of resources to obtain data that were readily available elsewhere.

Although none of the studies were aimed at training development, it was nevertheless felt that very poor use was made of training material, such as manuals. It is accepted that such material can tend to be biased towards the issues that are most difficult to learn, but nevertheless, it was considered that training information could have been useful for many of the studies, even where training was not a prime concern.

2.3. *Data collection and analysis*
HTA is perhaps the nearest thing to a universal TA technique. In fact, within the context of process control, TA is often wrongly assumed to be completely synonymous with HTA. It was thus interesting to discover that half of the studies did not undertake any HTA, but merely developed sequential task lists. While this can be appropriate for some scenario-based tasks, a hierarchical approach would certainly have been beneficial for others. Furthermore, even when HTAs were undertaken, HTA plans were only developed in a minority of cases. The lack of plans to show how complex task components were ordered within the HTAs was a serious shortcoming. Even in those tasks where the subordinate task components are obviously executed sequentially, it is incumbent upon the analyst to state this observation explicitly.

Many of the studies produced some form of decomposition table, although many referred to this process as tabular TA, rather than task decomposition. Generally, these decompositions were based upon the lowest level task descriptions and were very useful. However, in some cases, much of the information was only obtained for higher task levels and this was considered to be of limited utility. It was particularly interesting to note that the studies that used the most decomposition categories tended to produce the most detailed design recommendations. However, further study would be necessary to determine whether this was because some analysts were more thorough both in collecting data and making design recommendations, or whether the process of gathering more data and separating out different aspects of it, improved the analysts' understanding of potential problems.

Where subjective assessments of workload were made during a walk-through or while undertaking an actual task, a 5 point rating was judged to be slightly superior to the use of a 3-point scale, but both identified high workload elements. However, there was concern that asking participants to make workload assessments when they were only provided with written, often terse, task descriptions, may generate results that are neither accurate nor consistent. Therefore, some validation of this approach was felt to be essential. It was also disappointing to note that so much reliance was placed upon using workload analyses to determine functional allocations. While workload analyses can provide useful data for making functional allocation decisions, they are only one aspect that needs to be considered (Price 1985) and more innovative approaches are needed to include other factors, such as skill requirements, motivation and job content.

Graphical techniques, such as link analysis, operational sequence diagrams, flow charts and timelines were not always used effectively. Generally, this was because of the level of task complexity being represented, resulting in some diagrams being overly complicated, while others were used to show very trivial relationships. In other cases, the techniques were misused, for instance there was some confusion between link analysis and spatial operational sequence diagrams. There were also some notable examples of failures to exploit particular representations, either because the scales were inappropriate (especially with timelines), or because more advanced representations were not used. Therefore, it was concluded that more guidance was clearly necessary to improve the utilization of graphical techniques.

2.4. *Data presentation*

Although in the majority of reports some indication was given about the data sources and TA methods that had been used, these were sometimes very sketchy and only provided the briefest of outlines. Particular problems were:

(1) The analysts were not clearly identified in the report and so it was not possible to ascertain whether recommendations that were presented were based upon an understanding of the underlying psychological mechanisms, or whether they were merely subjective opinions.
(2) Where walk-throughs and other on-site data collection had been undertaken, it was often not clear how thorough this had been. For instance, the reader should be able to ascertain whether a report is based upon several site visits that have involved several different operators, or whether all the data were collected during a single, short visit.
(3) Subject matter experts were often used without defining their experience and background.
(4) The methods used for deriving task timings were often not reported.

It was also noted that while some reports provided detailed information about the analysis in the reports or in appendices, in other reports, this information was only given in a summarized form. Indeed, in some reports, even the underlying task descriptions, which are the main building blocks of any TA, were not presented. Without these background data, any reader or user of a TA will have difficulty in prioritizing the recommendations. Such difficulties will become more acute when it is not feasible to implement a particular recommendation in the way suggested by an analyst, because there will be little information available about the underlying problem to suggest alternative resolutions.

3. Nuclear industry survey

Within the nuclear industry over 60 TAs were identified after discussions with ergonomists employed in the industry. Cross-referencing from these TAs identified a further 10 reports. This list was considered to be representative of TAs undertaken within the industry, although it did not include some TAs that had been performed to obtain training information. TA reports for all but five of these TAs have been obtained and reviewed using a similar pro forma to that used for the survey of the defence industry.

It was clear from this review that although there was some variation in the quality of the different analyses, there was much more consistency in the analysis and reporting and the overall quality of the studies was considered to be generally higher than that observed in the defence survey. Follow-up discussions indicated that ergonomists employed within the nuclear industry were nearly always involved with the planning and monitoring of TAs. It was concluded that this factor was beneficial, because it ensured that the human issues in the projects were clearly defined at the outset and it provided a more prominent focus upon ergonomics and psychological concerns throughout the work.

3.1. *Scope of the studies*

Table 2 summarizes the scope of the survey. There were considerably more studies in this second survey, hence the reasons for the study have, in the case of interface design and performance assessment, been divided into sub-categories. In addition, it was felt pertinent to include in the table an indication of the way that the various studies related to particular phases in the system life cycle. The methods, data collection and presentation were generally the same as those used in the defence survey. These are shown in table 1 and thus this information has not been repeated.

Respondents to the survey were asked to categorize each study into one of the six key phases of the life cycle of the system under study. These phases were — concept, feasibility, project definition, detailed design, acceptance/commissioning and in service. The studies covered most stages of the design process, although TA appeared to be little used during the earliest design decisions. For instance, no TAs were reported during the concept stage while nine TAs overlapped the feasibility and project definition stages and thus the data for these two stages were combined.

This contrasts with the defence industry, where TA generally tends to be introduced earlier in the design process and logistical issues assume a greater importance. Although, to some extent this reflects differences between the two industries, it would be prudent for both industrial sectors to consider whether the general scope of their task analyses should be broadened within the design cycle.

3.2. *Data collection*

The three most popular means of data collection were talk-throughs, walk-throughs and informal discussions. These were often used together, but where it was possible to undertake a walk-through, even if this was only in a static mock-up, a preliminary talk-through only appeared to be useful as an orientation to analysis with little process knowledge.

Where a walk-through was possible, there appeared to be two different approaches. In some studies, analysts used a walk-through to develop a basic set of task descriptions and then more detailed information was obtained from informal discussions and other means. In others, analysts filled out comprehensive decomposition tables during their walk-throughs and then only undertook limited discussions to confirm the relationship between task elements or to resolve any

Table 2. Summarized results from studies undertaken in the nuclear industry.

Reason for study	Feasibility/ project definition	Detailed design	Acceptance/ commissioning	In service
Systems procurement	–	–	–	–
Allocation of function and workload assessment	2	5	–	9
Manpower analysis and staffing requirements	5	9	–	17
Interface design				
Hardware	6	11	9	11
Software	1	5	8	6
Procedures	4	9	11	14
Operability	7	11	11	24
Systems modelling	–	–	–	–
Training	3	8	8	8
Human error identification	4	5	–	20
Systems assessment				
Task timing	1	4	1	14
Human reliability	2	4	–	12
Performance assessment	2	5	1	9

outstanding issues. However, there was no clear evidence that either approach was preferable and so the analyst's personal preference is probably the best guide.

Only nine of the studies reported using table-top discussions between selected subject-matter experts. Although these appeared to provide useful data for the analyses, it was not possible to come to any conclusions about the optimal composition of such groups. However, all of these discussions were focused upon specific scenarios or procedures and this appeared to be a very effective format. Other analysts used questionnaires and structured interviews, but neither of these proved to be particularly effective ways to obtain task data.

The reviewers noted that only seven studies specifically referred to critical incidents. When questioned, many of the analysts indicated that this was because they had not found any critical incident reports that were particularly relevant. Thus, it was felt that it would be helpful to provide some mechanism for ensuring that analysts do make an effective search of critical incidents and that the search is reported even if no relevant information is uncovered.

3.3. *The use of HTA*

Although HTA was widely used, a considerable number of TAs used alternative techniques. In many of these cases HTA was not appropriate, either because the analyses involved relatively short task sequences, or because the analysis was limited to checking reasonably well-established written procedures. Nevertheless, there were 12 remaining TAs where the task descriptions appeared to be derived directly from the written procedures without any attempt at hierarchical redescription. Such an approach risks tying the analysis too closely to the existing procedures without checking whether this is justified. It also means that task steps may be missed, or else, if the tasks are defined at too high a level, important task features may not be checked. It was noted that within these particular TAs, the recommendations that were made tended to be relatively generic and lacked the psychological insights that were prominent in those studies that had applied HTA.

Most of the HTAs presented task plans and where these were not presented it appeared that the tasks would generally be undertaken as linear sequences. Nevertheless, it would have been helpful if this could have been specifically mentioned. The general preference was to present plans as text, but in some cases flow diagrams were used. For more complex plans, these diagrams were often unclear.

It was interesting to note that approximately one-half of the HTAs did not indicate any specific rules used by the analysts to determine when to halt the hierarchical redescription. Where the stopping rules were mentioned, a number of different criteria were given, for instance continuing to the level at which feedback or corrective action was possible, or continuing until a single control or instrument was used. However, regardless of the criteria that were applied, most of the HTAs used either four or five levels of redescription. Thus the reviewers concluded that for most of the studies, the substitution of different criteria would probably have made little difference to the actual stopping points. In other words, the analysts' *feel* for when to stop the redescription may well be as effective as the application of specific, defined criteria.

The presentation of HTA information in the written reports was very variable. One-third of the HTAs were not accompanied by HTA diagrams and some that were presented were very cluttered. It is considered that HTA diagrams fulfil two important functions: (1) they provide a structured overview of the task, and (2) they illustrate the model that has been used by the analyst in organizing the task components. HTA diagrams should, therefore be clear and comprehensible. Users who were unfamiliar with task analysis sometimes misinterpreted the HTA presentations and felt that it would be helpful to have an explanation of the various conventions that were used to present HTAs and associated information. However, this was only included in a minority of the investigated reports.

3.4. *Task decomposition*
Just over one-half of the studies used some form of task decomposition, although in many cases this was not formalized into a task decomposition table. However, it is likely that some of these were *post hoc* summaries of the data that had been collected, rather than systematic attempts to search for very specific types of data throughout all the task elements. As in the defence industry survey, it appeared that those analyses which were based upon about eight to ten decomposition categories also produced reports with more insightful recommendations, than those that used fewer.

3.5. *Other methods of data analysis and presentation*
The other most widely used method of data analysis was the production of timelines. This provided an effective method of demonstrating whether tasks were feasible. Nevertheless, timelines could have been presented more effectively, by using better scales, or by appropriate coding to emphasize particular information. In some cases, the minimum time segments were so coarse that the overall times became unrealistically pessimistic. It was also noted that in a minority of the studies the basis for the timings was not very clear and there was seldom any indication of underlying assumptions about ancillary tasks, such as communications, that determined the degree of pessimism, or otherwise, in the timings.

A dozen of the studies incorporated some form of error analysis, but the depth and scope of these analyses varied and there were insufficient data for any meaningful comparisons of different approaches. It was noted that decision-action diagrams were used very effectively to identify potential decision-making problems by those analysts with a psychological background.

3.6. *Depth of analyses*
Clearly it would be anticipated that the number and the nature of the recommendations that are given in a task analysis will depend upon the nature of the specific tasks that are being analysed. However, it was also very evident that some of the recommendations were based upon a much deeper understanding of the tasks than others. Therefore, a subjective assessment was made of the depth of recommendations in those analyses for which recommendations had been given. This indicated that 18 of the studies provided mainly generic recommendations, 33 provided detailed recommendations and a further five studies fell between the two categories.

A total of 25 of the studies that provided detailed recommendations also supported their specific recommendations with an explanation of the underlying problems. This was considered particularly useful because it enabled designers and others to develop effective alternative recommendations where there were practical problems with those proposed by the analyst. It was noted that all of the studies where detailed recommendations were accompanied by explanations were provided by ergonomists or psychologists. This again highlights the importance of ensuring that at least one of the analysts has appropriate human factors qualifications.

3.7. *Resourcing requirements for task analysis*

Very few of the reports provided any indication of the amount of resourcing that had been required to undertake the TA and when questioned, most of the analysts were unable to be very precise about the time that had been required for various aspects of the data collection and analysis. It was, however, evident that the analysts, who were judged to have produced the more in-depth TAs, spent correspondingly more time on site. Where most of the data were gathered off-site from reports and procedures, the TAs often appeared to overwhelm the readers with relatively superficial information that would have little impact upon task performance.

4.   Conclusions

Close to 100 TAs were examined in the two surveys. These provided some useful insights into the ways that TAs are undertaken and enabled the reviewers to identify areas where the TA process could be improved. Although some of the findings were very specific, generally there was much consistency between the two surveys and it is anticipated that many of the findings will be applicable to a wide range of TAs.

However, there was considerably more variability among the defence TAs compared to those undertaken in the nuclear industry, particularly in terms of their presentation and depth of analysis. Nevertheless, it should be noted that the best of the studies in both surveys were certainly comparable. Although the differences revealed in the two surveys could have been due to sampling, it was felt that, as the variation was so pronounced, alternative explanations should be considered. A detailed comparison of the underlying features of the TAs in the two surveys indicated that there were two factors that differed noticeably between the two industries. These were the TA objectives and the level of human factors expertise involved:

(1) In the defence industry TA objectives were not always presented very clearly to the analysts at the outset, whereas those responsible for commissioning task analyses within the nuclear industry tended to provide a very clear and comprehensive indication of the purposes of the TAs at the start of each project.

(2) The level of human factors expertise was generally higher amongst the analysts who undertook the task analyses in the nuclear industry survey.

It is suspected that the reduced variability and better overall quality in the nuclear industry TAs was largely the effect of a combination of these two factors.

At the outset, those persons responsible for commissioning TA projects need to define the objectives very clearly. They also need to have a feel for the effort involved in undertaking a thorough analysis, otherwise there is a real danger that they may avoid prescribing a comprehensive study by experienced analysts, in favour of a cheaper, but much more superficial study. In order to avoid these problems it is considered that there is a need to provide clear and systematic advice to those responsible for commissioning and managing TA exercises.

Overall, there was a marked tendency for analysts to be relatively parsimonious in reporting their methods and data sources. Thus, even for the most comprehensive of the studies, it was often difficult, or impossible, to determine those important details that are necessary to enable the users to assess the weight that should be placed upon any specific recommendation. This meant that they could be asked to accept potentially costly recommendations with little indication of the strength of the evidence.

Generally, in all the studies it appeared that the most insightful analyses were undertaken by analysts who had the most human factors experience. It became evident that task analyses carried out by staff who were not qualified ergonomists or psychologists tended to lack psychological insight and usually generated little more than task descriptions. Such analyses may be appropriate for some applications, but for complex tasks that impose high memory loads, require careful organization and planning, difficult decision-making or prediction, it appears that some psychological expertise is beneficial.

It was also evident that some analyses were undertaken to a substantially different depth and level of detail than others, but this was not always easy to ascertain. For example, one analysis of a substantial task transpired to have been based on only 1.5 days on site, compared with over 20 days on site for another, broadly similar, TA. However, this disparity only became apparent after subsequent interviews with the analysts concerned. The number of recommendations provides a clue to help judge the depth of analysis. The presence of few recommendations probably indicates a superficial analysis, although it may of course be the case that there were indeed few recommendations to be made. However, it must also be stressed that for many applications a less in-depth analysis may well be adequate.

A further item that could be useful in assessing the quality and depth of an analysis, is the number of decomposition categories utilized. In the studies that were reviewed, there was a clear relationship between the number of decomposition categories utilized and the depth of understanding displayed by the analysts. However, more research is needed to ascertain whether some analysts were more thorough both in collecting data and making design recommendations, or whether the process of gathering more data and separating out different aspects of it, did in itself lead to a better understanding of potential problems.

Essentially, it is apparent from the survey that there is a need for the development of guidelines to address the various stages of undertaking a TA, including planning the methodology, data collection, data analysis and the presentation of TA information. However, such guidelines should provide systematic advice, rather than inflexible, prescriptive rules.

Acknowledgements

The support of Technical Group 5 (Human Sciences and the Synthetic Environment) of the Ministry of Defence Corporate Research and of the Health and Safety Executive's Generic Nuclear Safety Research Programme is gratefully acknowledged. It should be noted that this paper presents the views and opinions of the authors. It should not be taken as a statement of either MoD or HSE policy.

References

AINSWORTH, L. and PENDLEBURY, G. 1995, Task-based contributions to the design and assessment of the man-machine interfaces for a pressurized water reactor, *Ergonomics*, **38**, 462–474.

PENDLEBURY, G. and AINSWORTH, L. 1992, Task analysis of operator safety actions, in B. Kirwan and L. Ainsworth (eds), *A Guide to Task Analysis*, (London: Taylor & Francis), 311–326.

PRICE, H. E. 1985, The allocation of functions in systems, *Human Factors*, **27**, 33–45.

CHAPTER 6

Applied cognitive task analysis (ACTA): a practitioner's toolkit for understanding cognitive task demands

Laura G. Militello and Robert J. B. Hutton

Klein Associates Inc., 582 E. Dayton-Yellow Springs Road, Fairborn, Ohio 43524, USA

Keywords: Cognitive task analysis; Knowledge elicitation; Knowledge representation; Expertise; Validity; Reliability.

Cognitive task analysis (CTA) is a set of methods for identifying cognitive skills, or mental demands, needed to perform a task proficiently. The product of the task analysis can be used to inform the design of interfaces and training systems. However, CTA is resource intensive and has previously been of limited use to design practitioners. A streamlined method of CTA, Applied Cognitive Task Analysis (ACTA), is presented in this paper. ACTA consists of three interview methods that help the practitioner to extract information about the cognitive demands and skills required for a task. ACTA also allows the practitioner to represent this information in a format that will translate more directly into applied products, such as improved training scenarios or interface recommendations. This paper will describe the three methods, an evaluation study conducted to assess the usability and usefulness of the methods, and some directions for future research for making cognitive task analysis accessible to practitioners. ACTA techniques were found to be easy to use, flexible, and to provide clear output. The information and training materials developed based on ACTA interviews were found to be accurate and important for training purposes.

1. Introduction

Task analytic techniques have played a critical role in the development of training and system design for the past 100 years starting with the pioneering work of Taylor (1911) and the Gilbreths (Gilbreth 1911, Gilbreth and Gilbreth 1919). (For more recent reviews of task analytic techniques see Drury *et al.* 1987, Kirwan and Ainsworth 1992). As task analytic techniques have become more sophisticated, focusing on cognitive activities as well as behaviours, they have become less accessible to practitioners. This paper introduces Applied Cognitive Task Analysis (ACTA), a set of streamlined cognitive task analysis tools that have been developed specifically for use by professionals who have not been trained in cognitive psychology, but who do develop applications that can benefit from the use of cognitive task analysis.

Cognitive task analysis methods focus on describing and representing the cognitive elements that underlie goal generation, decision making, judgements, etc. Although cognitive task analyses often begin with high-level descriptions of the task based on observations or initial interviews, the bulk of the data collection occurs via in-depth interviews with subject matter experts. These interviews focus on gaining information about the cognitive strategies used to accomplish the task including situation assessment strategies, identification and interpretation of critical cues, metacognitive strategies, important perceptual distinctions, etc. One strength of these methods is that they aid experts in articulating knowledge that is generally difficult to verbalize. Although researchers are often cautioned concerning reliance on verbal protocol data (Nisbett and Wilson 1977), it can be argued that verbal reports are inherently no more or less problematic than any other methodology (Howard 1994, Spector 1994).

The urgency of incorporating cognitive elements of performance into training and systems design stems from the changing nature of the workplace and the impact of technology on many tasks and functions. Tasks today place greatly increased demands on the cognitive skills of workers. Howell and Cooke (1989) have argued that with advances in technology and machine responsibility, we have increased, rather than lowered, cognitive demands on humans. More procedural or predictable tasks are handled by smart machines, while humans have become responsible for tasks that require inference, diagnoses, judgement, and decision making. In manufacturing technologies, for example, key skills now include perceptual skills for monitoring equipment, diagnostic skills for interpreting computerized information, and communication skills required for problem solving and co-ordination in distributed decision environments. These skills are typically grounded in hands-on experience with a task and are among the most difficult for proficient performers to access or articulate. Technology also has the potential to simplify low-level jobs; however, the concern shared by many is that it makes high-level jobs even more complex.

While a wide range of powerful methods of cognitive task analysis have been developed and applied over the last 10 years, few have become accessible to training practitioners and the engineering community designing systems. These have been described in several different sources (Klein 1993, Cooke 1994, Seamster *et al.* 1997). For example, in a multi-year effort Hall *et al.* (1994) used the Precursor, Action, Result, and Interpretation (PARI) method to develop an avionics troubleshooting tutor. Roth *et al.* (1991) used a cognitive environment simulation to investigate the cognitive activities involved in fault management with nuclear power plant operators. Seamster *et al.* (1993) report conducting an extensive cognitive task analysis to specify the instructional content and sequencing for a US Federal Aviation Administration *en route* air traffic control curriculum. Rasmussen (1986) has conducted a thorough cognitive analysis of nuclear power plant control room operation. Rouse (1984) has used similar methods to derive problem-solving strategies for troubleshooting tasks in the military.

Although these success stories are very persuasive in terms of the power of cognitive task analysis, all have required considerable time and resources. All have been part of research efforts conducted by scientists as opposed to the development of an application by practitioners. The objective of the authors has been to transition CTA techniques from the research community to the operational community. Based on existing CTA techniques, the authors have developed streamlined CTA methods intended for use by instructional designers and systems designers rather than knowledge engineers, cognitive psychologists, and human factors/ergonomics professionals. This paper describes Applied Cognitive Task Analysis (ACTA), streamlined CTA methods developed for training practitioners and systems designers to elicit and represent cognitive components of skilled task performance, and the means to transform those data into design recommendations.

2. Applied Cognitive Task Analysis

The ACTA techniques were developed as part of a two-year project funded by the Navy Personnel Research and Development Center. The goal of this project was to develop and evaluate techniques that would enable instructional designers and systems designers to elicit critical cognitive elements from Subject Matter Experts (SMEs). The techniques presented here are intended to be complementary; each is designed to get at different aspects of cognitive skill. The first technique, the task diagram interview, provides the interviewer with a broad overview of the task and highlights the difficult cognitive portions of the task to be probed further with in-depth interviews. The second technique, the knowledge audit, surveys the aspects of expertise required for a specific task or subtask. As each aspect of expertise is uncovered, it is probed for concrete examples in the context of the job, cues and strategies used, and why it presents a challenge to inexperienced people. The third technique, the simulation interview, allows the interviewer to probe the cognitive processes of the SMEs within the context of a specific scenario. The use of a simulation or scenario provides job context that is difficult to obtain via the other interview techniques, and therefore allows additional probing around issues such as situation assessment, how situation assessment impacts a course of action, and potential errors that a novice would be likely to make given the same situation. Finally, a cognitive demands table is offered as a means to consolidate and synthesize the data, so that it can be directly applied to a specific project. Each technique is described in detail in the following sections.

2.1. *Task diagram*

The task diagram elicits a broad overview of the task and identifies the difficult cognitive elements. Although this preliminary interview offers only a surface-level view of the cognitive elements of the task, it enables the interviewer to focus the more in-depth interviews (i.e. the knowledge audit and simulation interviews) so that time and resources can be spent unpacking the most difficult and relevant of those cognitive elements.

The subject matter expert is asked to decompose the task into steps or subtasks with a question such as, 'Think about what you do when you (*task of interest*). Can you break this task down into less than six, but more than three steps?' The goal is to get the expert to walk through the task in his/her mind, verbalizing major steps. The interviewer limits the SME to between three and six steps, to ensure that time is not wasted delving into minute detail during the surface-level interview. After the steps of the task have been articulated, the SME is asked to identify which of the steps require cognitive skill, with a question such as, 'Of the steps you have just identified which require difficult cognitive skills? By cognitive skills I mean judgements, assessments, problem solving – thinking skills'. The resulting diagram (figure 1) serves as a road map for future interviews, providing an overview of the major steps involved in the task and the sequence in which the steps are carried out, as well as which of the steps require the most cognitive skill.

The task diagram interview is intended to provide a surface-level look at the task, and does not attempt to unpack the mental model of each SME. The goal is to elicit a very broad overview of the task. Efforts to delineate a mental model can quickly degenerate into a search for everything in a person's head as Rouse and Morris (1986) have pointed out. In this interview, the authors recommend that the SME be limited to six steps and advise the interviewer not to get dragged down to a level of detail that is best captured by other interview techniques.

2.2. *Knowledge audit*

The knowledge audit identifies ways in which expertise is used in a domain and provides examples based on actual experience. The knowledge audit draws directly from the research literature on expert-novice differences (Dreyfus 1972, Chi *et al.* 1981, Shanteau 1985, Dreyfus and Dreyfus 1986, Hoffman 1992, Klein and Hoffman 1993) and critical decision method studies (Klein *et al.* 1989, Crandall and Getchell-Reiter 1993, Klinger and Gomes 1993, Militello and Lim 1995, Kaempf *et al.* 1996) of expert decision-making. The knowledge audit has been developed as a means for capturing the most important aspects of expertise while streamlining the intensive data collection and analysis methods that typify studies of expertise.

The knowledge audit is organized around knowledge categories that have been found to characterize expertise. These include; diagnosing and predicting, situation awareness, perceptual skills, developing and knowing when to apply tricks of the trade, improvising, metacognition, recognizing anomalies, and compensating for equipment limitations. Clearly, the authors could have included many more items, but the intent was to aim for the smallest number of high impact components.

The knowledge audit employs a set of probes designed to describe types of domain knowledge or skill and elicit appropriate examples (figure 2). The goal is not simply to find out whether each component is present in the task, but to find out the nature of these skills, specific events where they were required, strategies that have been used, and so forth. The list of probes is the starting point for conducting this interview. Then, the interviewer asks for specifics about the example in terms of critical cues and strategies of decision making. This is followed by a discussion of potential errors that a novice, less-experienced person might have made in this situation.

The examples elicited with the knowledge audit do not contain the extensive detail and sense of dynamics that more labour-intensive methods such as the critical decision method (Klein *et al.* 1989) incident accounts often do. However, they do provide enough detail to retain the appropriate context of the incident. It is not expected that all probes will be equally relevant in each domain. After a few interviews, interviewers can easily determine which probes have the highest payoff. Although the knowledge audit does not capture the depth of relationship of a conceptual graph structure (Gordon and Gill 1992) or other intensive methods, it does address a full range of aspects of expertise that are usually neglected by behavioural task analytic methods.

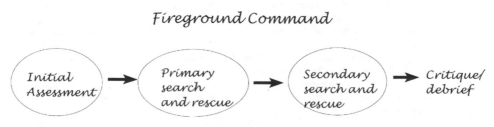

Figure 1.   Task diagram example.

The output of the knowledge audit is shown in table 1, which contains an inventory of task-specific expertise. This table includes examples of situations in which experience has been called into play, cues and strategies used in dealing with these difficult situations, and an explanation of why such situations present a challenge to less-experienced operators.

2.3. *Simulation interview*

The simulation interview allows the interviewer to better understand the SME's cognitive processes within the context of an incident. In operational settings, the point of the job is typically to act upon the environment in some manner. Klein (1993) and

---

BASIC PROBES:

☐ **Past & Future**. Experts can figure out how a situation developed, and they can think into the future to see where the situation is going. Among other things, this can allow experts to head off problems before they develop.

> *Is there a time when you walked into the middle of a situation and knew exactly how things got there and where they were headed?*

☐ **Big Picture**. Novices may only see bits and pieces. Experts are able to quickly build an understanding of the whole situation—the Big Picture view. This allows the expert to think about how different elements fit together and affect each other.

> *Can you give me an example of what is important about the Big Picture for this task? What are the major elements you have to know and keep track of?*

☐ **Noticing.** Experts are able to detect cues and see meaningful patterns that less-experienced personnel may miss altogether.

> *Have you had experiences where part of a situation just "popped' out at you; where you noticed things going on that others didn't catch? What is an example?*

☐ **Job Smarts.** Experts learn how to combine procedures and work the task in the most efficient way possible. They don't cut corners, but they don't waste time and resources either.

> *When you do this task, are there ways of working smart or accomplishing more with less—that you have found especially useful?*

☐ **Opportunities/Improvising.** Experts are comfortable improvising—seeing what will work in this particular situation; they are able to shift directions to take advantage of opportunities.

> *Can you think of an example when you have improvised in this task or noticed an opportunity to do something better?*

☐ **Self Monitoring.** Experts are aware of their performance; they check how they are doing and make adjustments. Experts notice when their performance is not what it should be (this could be due to stress, fatigue, high workload, etc.) and are able to adjust so that the job gets done.

> *Can you think of a time when you realized that you would need to change the way you were performing in order to get the job done?*

OPTIONAL PROBES:

☐ **Anomalies.** Novices don't know what is typical, so they have a hard time identifying what is atypical. Experts can quickly spot unusual events and detect deviations. And, they are able to notice when something that ought to happen, doesn't.

> *Can you describe an instance when you spotted a deviation from the norm, or knew something was amiss?*

☐ **Equipment Difficulties.** Equipment can sometimes mislead. Novices usually believe whatever the equipment tells them; they don't know when to be skeptical.

> *Have there been times when the equipment pointed in one direction, but your own judgment told you to do something else? Or when you had to rely on experience to avoid being led astray by the equipment?*

Figure 2.   Knowledge audit probes.

Howell and Cooke (1989) have asserted that identification and explor
information surrounding high consequence, difficult decisions can provide
basis for generation of effective training and systems design. Simulatic
incident-based interviews have been used successfully in many domains (Fl
1954, Grover 1983, Clarke 1987, Diederich *et al.* 1987, Bell and Hardiman 1989,
Cordingley 1989, Klein *et al.* 1989, Thordsen *et al.* 1992, Hall *et al.* 1994).

The simulation interview is based on presentation of a challenging scenario to the
SME. The authors recommend that the interviewer retrieves a scenario that already
exists for use in this interview. Often, simulations and scenarios exist for training
purposes. It may be necessary to adapt or modify the scenario to conform to
practical constraints such as time limitations. Developing a new simulation
specifically for use in the interview is not a trivial task and is likely to require an
upfront CTA in order to gather the foundational information needed to present a
challenging situation. The simulation can be in the form of a paper-and-pencil
exercise, perhaps using maps or other diagrams. In some settings it may be possible
to use video or computer-supported simulations. Surprisingly, in the authors'
experience, the fidelity of the simulation is not an important issue. The key is that the
simulation presents a challenging scenario.

After exposure to the simulation, the SME is asked to identify major events,
including judgements and decisions, with a question such as, 'As you experience this
simulation, imagine you are the (*job you are investigating*) in the incident.
Afterwards, I am going to ask you a series of questions about how you would
think and act in this situation'. Each event is probed for situation assessment,
actions, critical cues, and potential errors surrounding that event (figure 3).

Table 1. Example of a knowledge audit table.

Aspects of expertise	Cues and strategies	Why difficult?
Past and future e.g. Explosion in office strip; search the office areas rather than source of explosion	Material safety data sheets (MSDS) tell you that explosion in area of dangerous chemicals and information about chemicals Start where most likely to find victims and own safety considerations	Novice would be trained to start at source and work out May not look at MSDS, to find potential source of explosion, and account for where people are most likely to be
Big picture Big picture includes source of hazard, potential location of victims, ingress/egress routes, other hazards	Senses, communication with others, building owners, MSDS, building pre-plans	Novice gets tunnel vision, focuses on one thing, e.g. victims
Noticing Breathing sounds of victims	Both you and partner stop, hold your breath, and listen Listen for crying, talking to themselves, victims knocking things over	Noise from own breathing in apparatus, fire noises Don't know what kinds of sounds to listen for

Information elicited is recorded in the simulation interview table (table 2). Using the same simulation for interviews with multiple SMEs can provide insight into situations in which more than one action would be acceptable, and alternative assessments of the same situation are plausible. This technique can be used to highlight differing SME perspectives, which is important information for developing training and system design recommendations. The technique can also be used to contrast expert and novice perspectives by conducting interviews with people of differing levels of expertise using the same simulation.

For each major event, elicit the following information

    As the (job you are investigating) in this scenario, what actions, if any, would you take at this point in time?

    What do you think is going on here? What is your assessment of the situation at this point in time?

    What pieces of information led you to this situation assessment and these actions?

    What errors would an inexperienced person be likely to make in this situation?

Figure 3.   Simulation interview probes.

Table 2. Example of a simulation interview table.

Events	Actions	Assessment	Critical cues	Potential errors
On-scene arrival	Account for people (names) Ask neighbours (but don't take their word for it, check it out yourself) Must knock on or knock down to make sure people aren't there	It's a cold night, need to find place for people who have been evacuated	Night time Cold –> 15° Dead space Add on floor Poor materials wood (punk board), metal girders (buckle and break under fire) Common attic in whole building	Not keeping track of people (could be looking for people who are not there)
Initial attack	Watch for signs of building collapse If signs of building collapse, evacuate and throw water on it from outside	Faulty construction, building may collapse	Signs of building collapse include: What walls are doing: cracking What floors are doing: groaning What metal girders are doing: clicking, popping Cable in old buildings hold walls together	Ventilating the attic, this draws the fire up and spreads it through the pipes and electrical system

2.4 *Cognitive demands table*

After conducting ACTA interviews with multiple SMEs, the authors recommend the use of a cognitive demands table (table 3) to sort through and analyse the data. Clearly, not every bit of information discussed in an interview will be relevant for the goals of a specific project. The cognitive demands table is intended to provide a format for the practitioner to use in focusing the analysis on project goals. The authors offer sample headings for the table based on analyses that they have conducted in the past (difficult cognitive element, why difficult, common errors, and cues and strategies used), but recommend that practitioners focus on the types of information that they will need to develop a new course or design a new system. The table also helps the practitioner see common themes in the data, as well as conflicting information given by multiple SMEs.

3. Evaluation study

As Hoffman *et al.* (1998) point out, the question of how to empirically verify a knowledge base, and the methodologies used to articulate and represent that knowledge base, has received little attention from the research community. Many

Table 3. Example of a cognitive demands table.

Difficult cognitive element	Why difficult?	Common errors	Cues and strategies used
Knowing where to search after an explosion	Novices may not be trained in dealing with explosions. Other training suggests you should start at the source and work outward Not everyone knows about the Material Safety Data Sheets. These contain critical information	Novice would be likely to start at the source of the explosion. Starting at the source is a rule of thumb for most other kinds of incidents	Start where you are most likely to find victims, keeping in mind safety considerations Refer to Material Safety Data Sheets to determine where dangerous chemicals are likely to be Consider the type of structure and where victims are likely to be Consider the likelihood of further explosions. Keep in mind the safety of your crew
Finding victims in a burning building	There are lots of distracting noises. If you are nervous or tired, your own breathing makes it hard to hear anything else	Novices sometimes don't recognize their own breathing sounds; they mistakenly think they hear a victim breathing	Both you and your partner stop, hold your breath, and listen Listen for crying, victims talking to themselves, victims knocking things over, etc.

cognitive task analysis methods are evaluated solely on the basis of subjective judgements of whether or not they seemed to work for a particular application or project. Exceptions include work by Crandall and her colleagues in assessing the validity and reliability of data-gathering skills using the critical decision method (Taynor *et al.* 1987, Crandall and Calderwood 1989, Crandall and Gamblian 1991, Crandall and Getchell-Reiter 1993), and method comparisons by Hoffman and his colleagues (Hoffman 1987, Hoffman *et al.* 1995). The evaluation study described in this section not only attempts to address issues of validity and reliability for a specific set of CTA techniques, but also addresses a number of issues that surround the assessment of validity and reliability within the context of real-world tasks.

3.1. *Methods*

An evaluation of the ACTA techniques was conducted to establish the validity and reliability of the data gathered using the methods, and to assess the usability of the techniques. In addition, a comparison of information gathered using ACTA techniques to data gathered using unstructured interview techniques was conducted.

Parallel studies were conducted in two domains for this evaluation. The authors' intention was to test the ACTA techniques with a sample of naïve users – people who lacked knowledge or experience with cognitive task analysis or instructional design. A novice sample would allow a cleaner examination of the impact of the ACTA methods on the kind and quality of data produced. Therefore, students from graduate programmes in clinical, human factors, or cognitive psychology were recruited via postings on college bulletin boards and e-mail, to conduct interviews and generate instructional materials in either the firefighting domain or the Electronic Warfare (EW) domain. Volunteers were screened to make sure that they had no previous knowledge of the domain they would be investigating, no previous experience in conducting CTA, and no extensive experience or training in developing course materials. Each student was paid US $250 for participation in the project. Twelve students conducted interviews in the firefighting domain and 11 in the EW domain. The SMEs interviewed were experienced Fireground Commanders from the greater Dayton area in Dayton, Ohio, USA, who had at least 10 years of experience and were recommended by the Fire Chief of each local fire department; and experienced EW technicians from Fleet Training Center Pacific in San Diego, California, USA, who had at least 6 years of experience as EW technicians, including 4 years at sea and experience as an EW supervisor.

Within each domain, students were placed in one of two groups. An attempt was made to match the groups by age, gender, and educational level. After matching the students on these criteria, they were randomly assigned to groups. All students attended a 2-h workshop introducing the concepts underlying cognitive task analysis, describing the application of cognitive task analysis to the development of instructional materials, and providing a brief overview of the domain and specific task that they would be investigating. They also received instruction regarding the training materials they would be asked to develop following their interviews with SMEs. These materials included a cognitive demands table, learning objectives, and modifications to a training manual.

After the initial 2-h workshop, the matched groups of students were separated for the remainder of the study. One group, referred to as the Unstructured group, was provided with instructions to conduct interviews with SMEs in whatever format they believed would be most useful for gathering cognitive information. They were told to spend time preparing questions, but were not given any direction regarding how to structure the interviews or specific types of questions to ask. The other group, referred to as the ACTA group, was provided with a 6-h workshop on the ACTA techniques, including knowledge elicitation and knowledge representation.

Students in both the Unstructured and ACTA groups then participated in two interviews with SMEs. Each student led one interview with an SME and observed an interview conducted by another student with a different SME, and thus had access to data from two interviews. Each student in the ACTA group led one ACTA interview and observed an ACTA interview in the same domain; each student in the Unstructured group led one unstructured interview and observed one unstructured interview in the same domain. No SME was interviewed twice. Students working in the firefighting domain were asked to focus on the size-up task. Students working in the EW domain were asked to focus on signal threat analysis. All interviews were scheduled for a 3-h block of time.

Within a week of completing the two interviews, each student attended a 4-h session to analyse the data and develop training materials. The students were instructed not to collaborate or do any additional work with their interview notes prior to the scheduled session. During the 4-h session, they were required to structure and represent the information obtained in interviews. They were provided with materials and instructions and asked to:

(1) consolidate the data from the interview using the cognitive demands table format;
(2) develop at least ten cognitive learning objectives for a hypothetical course in that domain;
(3) revise or add to training manuals (these were provided), based on what they had learned in the interviews; and
(4) complete a questionnaire about participation in the study.

In addition, all students who had been exposed to the ACTA techniques were asked to fill out an ACTA usability questionnaire.

3.2. *Data transformation*

In order to generate quantitative measure of utility and validity, the information generated by the ACTA tools required extensive data codification and transformation. All materials generated by the sample of graduate students were assessed by SMEs who had not participated in the study thus far and/or cognitive psychologists. Wherever possible, data evaluation was carried out by multiple coders, so that inter-rater reliability could be assessed. In some cases, owing to lack of availability or resource constraints, only one SME was available to code the data. Measures were devised to address two aspects of validity: (1) whether the ACTA tools produced information that was predominantly cognitive in nature; and (2) whether the information produced was domain-specific and relevant. Data transformation procedures and associated measures are described in detail in the following sections.

3.2.1. *Validity indices—cognitive demands tables*: All items included in the cognitive demands tables were coded by two Klein Associates researchers, Laura Militello and Dr Rebecca Pliske, blind to treatment group (ACTA versus Unstructured), for whether they contained cognitive content. The criterion for inclusion in the cognitive demand category was that the item addressed a cognitive skill or a cognitive challenge that a firefighter/EW operator encounters on the job (e.g. 'deciding whether or not water supply on the scene will be adequate', 'devising a strategy to successfully remove people from a burning building'). Items in the non-cognitive demands category typically referred to declarative knowledge that the firefighter/EW operator should have (e.g. 'know the initial command sequence') or behaviours (e.g. 'return resources'). Although it could be argued that knowing the proper command sequence has a cognitive component, it does not constitute a cognitive challenge or skill needed to serve as a proficient Fireground Commander. Rather, it represents a type of background knowledge that one must obtain before becoming a Fireground Commander.

In order to establish whether students using the ACTA techniques could consistently elicit information across relevant cognitive categories (as opposed to task-based categories), the authors developed a coding scheme based on Rasmussen's model of decision making (Rasmussen 1986, Rasmussen *et al.* 1994). The categories included information collection, situation analysis, diagnosis, prediction, value judgement, choice, planning, and scheduling. Two raters, blind to the students' interview group (ACTA versus Unstructured), independently rated 30% of the data. The raters established acceptable inter-rater agreement (percentage agreement = 74%). Although no standard for acceptable inter-rater agreement exists (Meister 1985, Pedhazur and Schmelkin 1991), agreement ratings exceeding 70% are generally accepted as adequate for this type of coding. Subsets of data were analysed until an acceptable level of agreement was reached. The remaining data were then analysed by one coder. The rest of the data were then divided among the two raters to complete the rating.

Evaluation of the domain-specific content of the cognitive demands tables was based on the firefighting portion of the database. A task-based coding scheme specific to the firefighting domain was developed. Based on firefighting manuals made available to the authors by the National Fire Academy, Emmitsburg, MD, USA, it was established that there are three primary subtasks for which the Fireground Commander is responsible: size-up, strategy/tactics, and management. For the firefighter data, the coders independently assessed the content of each cognitive demands table item and assigned it to one of these three categories. The coders established reliability (percentage agreement = 81%) on 40% of the data, and one researcher then coded the remainder of the data.

The authors believed that it was also important to have the data evaluated by domain experts, in order to assess data quality and relevance. An assessment of the firefighter data was carried out by an instructor for the Incident Command course at the Ohio Fire Academy. He had more than 10 years of firefighting experience, had served as a Fireground Commander, and is currently involved in the development of course materials for the firefighter courses taught at the Ohio Fire Academy. The EW SME was a retired US Navy Electronic Warfare technician who had extensive experience as an operator, a supervisor, and as an instructor. The SMEs were asked to indicate what percentage of the information contained in each cognitive demands table would be likely to be known only by experienced personnel. In addition, the SMEs were asked to indicate the percentage of information contained in each cognitive demands table that would be relevant for experienced, highly-skilled personnel (Fireground Commander/EW supervisor) as opposed to a person with little experience on the job (firefighter/new EW operator). Given that one objective of the ACTA techniques is to elicit experience-based knowledge (as opposed to classroom knowledge, which is easily captured using other traditional techniques and disseminated via textbooks), the authors wanted to distinguish information that only an experienced person would know from that which people newly released from training would know. Both of these questions were intended to distinguish between information reflective of experience-based knowledge versus classroom knowledge.

3.2.2. Validity indices—instructional materials: In addition to the firefighter SME described above, a second instructor from the Ohio Fire Academy was recruited to provide ratings of the instructional materials generated by the students in the firefighting domain. He also had more than 10 years experience as a firefighter, had served as a Fireground Commander, and is currently involved in the development of course materials at the Ohio Fire Academy. In addition, two EW instructors from the Electronic Warfare 'A' School in Pensacola, Florida, USA, were recruited to rate the instructional materials generated by graduate students working in the EW domain. Both Electronic Warfare SMEs held a rank of E6 or above, had served as an Electronic Warfare supervisor, and had experience as an instructor at the EW 'A' School.

Working independently, the SMEs in each domain were asked to evaluate each learning objective and training manual modification for accuracy, importance, and whether or not it was currently included in the typical firefighter training/EW instructional programme. Validity ratings were made on a 3-point scale where 1 = not important, 2 = important, and 3 = very important. Accuracy ratings were made on a two-point scale where 1 = not accurate and 2 = accurate. In the firefighting domain, acceptable inter-rater agreement was obtained for the accuracy and importance ratings, but not for the rating of whether or not the information described in the learning objective was currently covered in the typical firefighter training course. Owing to unacceptable reliability ratings, no further analyses were conducted on data relating to whether the information was currently covered in a course. Discussion with the firefighter SMEs revealed that they had experience in teaching different courses and therefore had different perspectives on what was 'typically' included in firefighter instruction. For the learning objectives, the percentage agreement for the accuracy judgements of the firefighter SMEs was 87.8%; the percentage agreement for the importance ratings of the firefighter SMEs was 71.4%. For the modifications to the student manual, percentage agreement for accuracy was 90.1% and for importance it was 76.1%. The accuracy and importance ratings made by the SME who had more extensive experience in developing training materials for Fireground Commanders were used in further analyses.

The SMEs in the EW domain were not able to reach acceptable inter-rater agreement. For the learning objectives, percentage agreement for importance was 58.5% and for accuracy it was 67.9%. For the modifications to the student manual, percentage agreement for importance was 34.2% and for accuracy it was 61.7%. Discussion with the SMEs revealed that depending upon the type of ship one serves on, the EW job may be very different. The two SMEs had served on different types of ships and were currently teaching very different courses (basic tactics versus introductory course on equipment). As a result, they had quite different perspectives on what was important for an EW operator to learn in school to prepare him/her for the job. For all further analyses, the authors used the ratings from the SME with the most recent and most extensive sea experience.

3.3 *Results*

The results section first presents the authors' findings as they relate to the usability, validity, and reliability of the ACTA techniques, as these were the primary questions to be answered by the evaluation study. Thus the data presented in the following sections are based only on students who completed the ACTA workshops and used these methods to conduct an interview with an SME. The final portion of the results section discusses the data as they relate to differences between the materials generated by students who conducted interviews using ACTA versus those students who conducted unstructured cognitive interviews. Although few group differences were found, a discussion of how large intra-group variability impacted on this study is presented.

3.3.1. *Usability*:

In evaluating the ACTA tools, the authors were interested in understanding the subjective experiences of both the interviewers and the interviewees. User acceptance was key to the success of this project. To assess user acceptance, three questionnaires were administered: a usability questionnaire focusing specifically on the ACTA techniques, an interviewee questionnaire eliciting information from the SME's perspective, and an interviewer questionnaire addressing the experience of participating in all aspects of the study. The findings from the questionnaire data are presented next.

3.3.1.1. *Usability questionnaire*:

A usability questionnaire was administered to all graduate students who used the ACTA techniques. Overall, ratings were very positive. All of the tools were rated as useful. A 3-way, mixed design ANOVA taking into account the domain (Firefighting, Electronic Warfare), the ACTA techniques (Task Diagram, Knowledge Audit, Simulation Interview), and the individual questions (Questions 2, 3, 4, and 5) on the usability questionnaire showed no difference in the usability of the three techniques, $F(2, 18) = 1.34$, $p = 0.29$, or in the usability of the techniques across domains, $F(1, 9)$, $p < 1$. Mean ratings on all dimensions were above '3' on a 5-point scale, where '5' is the most positive rating and '1' is the least positive rating (table 4). These data indicate that graduate students found:

(1) the methods to be easy to use;
(2) the interview guides and job aids to be flexible;
(3) the output of the interviews to be clear; and
(4) the knowledge representations to be useful.

3.3.1.2. *Interviewee questionnaire*:

Each SME was asked to fill out a brief questionnaire at the end of the interview. If the ACTA tools are to be accepted in an operational community, the impressions of the people who are interviewed will have considerable influence. If the SMEs find the interview process aversive, or do not find that they are given an opportunity to communicate the critical elements of the job, acceptance of the ACTA tools will be greatly compromised within an organization.

The questionnaire data indicate that the interviewees found the interview experience to be pleasant and worthwhile. Table 5 presents the means for each question for those SMEs who participated in ACTA interviews. A 3-way, mixed design ANOVA taking into account domain (Firefighting, Electronic Warfare), interview type (ACTA, Unstructured), and question (5 questions from question-

Table 4. Usability questionnaire means of the graduate students who conducted interviews using the ACTA techniques.

Questions	Task diagram		Knowledge audit		Simulation interview		ACTA overall	
	Fire-fighting ($n=6$)	Electronic warfare ($n=5$)	Fire-fighting ($n=6$)	Electronic warfare ($n=5$)	Fire-fighting ($n=6$)	Electronic warfare ($n=5$)	Fire-fighting ($n=6$)	Electronic warfare ($n=5$)
Circle the amount of time you spent becoming familar with the tool before using it†	1.00 (0.00)	1.40 (0.55)	1.17 (0.41)	1.40 (0.55)	1.00 (0.00)	1.20 (0.45)	NA	NA
Rate the degree to which you found this technique easy to use	4.67 (0.52)	4.60 (0.89)	3.67 (1.03)	3.80 (1.30)	3.83 (1.60)	4.40 (0.89)	4.17 (0.75)	4.20 (0.45)
Rate the degree to which you found the interview guide to be flexible	4.17 (0.75)	3.40 (0.55)	3.67 (1.03)	3.40 (0.55)	3.83 (1.17)	3.60 (1.14)	3.67 (1.21)	3.80 (0.45)
Rate the degree to which you found the output to be clear	4.67 (0.52)	4.00 (1.00)	4.00 (1.10)	3.60 (0.55)	3.83 (1.83)	4.60 (0.55)	4.00 (1.10)	4.00 (0.00)
Rate the degree to which you found the knowledge representation to be useful	4.50 (0.55)	4.00 (0.71)	4.17 (1.17)	3.80 (0.84)	3.33 (1.21)	4.00 (0.71)	4.33 (1.21)	4.40 (0.55)

*All questions used a 5-point rating scale where 5 is the most positive rating and 1 is the least positive rating, unless otherwise stated. Standard deviations are included in parentheses.
†A 3-point scale where $1 = 1$ h or less; $2 = 1$ to 2 h; and $3 =$ more than 2 h was used for the questions concerning time required to prepare for each tool.

Table 5. Interviewee questionnaire means for the SMEs interviewed using the ACTA techniques.

Questions	Firefighting ($n=6$)	Electronic warfare ($n=5$)	Mean totals ($n=11$)
Overall, I found the interview to be a pleasant experience	4.33 (0.52)	3.60 (1.14)	3.97 (0.89)
The format of the interview allowed me to describe my expertise	4.50 (0.55)	3.60 (1.14)	4.05 (0.94)
I thought the interview lasted too long† (1 = strongly agree, 5 = strongly disagree)	4.00 (0.63)	4.00 (0.71)	4.00 (0.63)
Participating in the interview gave me new insights into the cognitive aspects of my job	3.67 (1.03)	3.20 (1.48)	3.44 (1.21)
I think the cognitive aspects of my job that were discussed during the interview are important things for a novice to learn	4.33 (0.52)	4.00 (1.22)	4.17 (0.87)

*All questions used a 5-point rating scale. Unless otherwise stated, 1 = strongly disagree, 5 = strongly agree. Standard deviations are included in parentheses.
†The scale for question 3 was switched so that a high score indicates a more positive response to be consistent with other questions on the questionnaire.

naire) showed no domain differences in the interviewees' responses, $F(1, 20) = 0.82$, $p < 1$.

3.3.1.3. *Interviewer questionnaire*: All graduate students filled out a questionnaire at the completion of their participation in the evaluation study. This questionnaire consisted of 10 questions intended to capture the interviewer's subjective experience, addressing issues such as confidence level, perceived difficulty or ease of the cognitive task analysis process, etc. The means for each question from the ACTA group are presented in table 6.

These data indicate that graduate students in both domains found the interviews to be informative and to provide cognitive information about the job domain. Based on the information learned via ACTA interviews, the graduate students found the development of a cognitive demands table and the generation of learning objectives to be easy. Participants indicated that they were able to make important revisions to the course materials provided. Surprisingly, participants responded negatively to the statement, 'I want to conduct more interviews because I still want more information'. The authors' only explanation for this is that because the participants

Table 6. Interviewer questionnaire means for the graduate students who conducted interviews using the ACTA techniques.

Questions*	Firefighting ($n=6$)	Electronic warfare ($n=5$)	Means total ($n=11$)
I felt confident in my ability to lead an interview	4.33 (0.82)	4.00 (0.00)	4.17 (0.60)
I learned more information from the interview I observed, than from the one I led	2.00 (0.89)	3.00 (1.23)	2.50 (1.13)
I felt I had sufficient information to revise the course materials	4.00 (0.89)	3.20 (1.30)	3.60 (1.12)
The interviews provided me with important information about the cognitive skills involved in this job domain	4.33 (0.52)	3.60 (1.14)	3.97 (0.89)
I wanted to conduct more interviews because I still wanted more information	2.33 (0.82)	3.20 (1.30)	2.77 (1.10)
The cognitive demands table was easy to fill out	4.00 (1.26)	3.40 (0.89)	3.70 (1.10)
It was easy to develop course objectives based on information specified in the cognitive demands table	4.33 (0.52)	3.20 (1.64)	3.77 (1.25)
I was able to use the information to make important changes in the course material	3.50 (0.84)	3.40 (1.14)	3.45 (0.93)
Overall, I found the interviews to be informative	4.83 (0.41)	4.60 (0.55)	4.72 (0.47)
Given the information, I found the revision of course materials straightforward	3.67 (0.82)	3.00 (1.23)	3.34 (1.03)

*All questions used a 5-point rating scale. Unless otherwise stated, 1 = strongly disagree, 5 = strongly agree. Standard deviations are included in parentheses.

were novices to cognitive task analysis, they did not anticipate the breadth and depth of knowledge that can be gained via these techniques.

3.3.2. *Validity*: Table 7 presents data that addresses three central questions regarding validity:

(1) Does the information gathered address cognitive issues?
(2) Does the information gathered deal with experience-based knowledge as opposed to classroom-based knowledge?
(3) Do the instructional materials generated contain accurate information that is important for novices to learn?

The extent to which the information elicited using ACTA was cognitive in nature was assessed by examining every item contained in the cognitive demands tables for its cognitive content. The cognitive content codings indicate that fully 93% of the items generated address cognitive issues. More specifically, in the firefighter study, 92% of the items were rated as cognitive and in the EW study 94% of the cognitive demand items generated by the students using ACTA were rated as cognitive.

To address the issue of whether the ACTA tool provided a means to elicit experience-based knowledge, SMEs were asked to make a global assessment of each cognitive demands table and to assign a percentage to each that reflected the proportion of information it contained that only highly-experienced personnel would be likely to know. The inference here is that such information is reflective of experience-

Table 7. Quality of outputs for graduate students who conducted interviews using the ACTA techniques.

Validity indicator	Firefighter ($n=6$)	Electronic warfare ($n=5$)
Percentage of total cognitive demands table items coded as cognitive	92%	94%
Proportion of cognitive demands table information experienced personnel likely to know, averaged across ACTA users	0.95 (0.05)	0.90 (0.09)
Proportion of cognitive demands table information relevant to a Fireground Commander/Electronic Warfare Supervisor, averaged across ACTA users	0.73 (0.10)	0.87 (0.10)
Proportion of student manual modifications rated as important or somewhat important, averaged across ACTA users	0.70 (0.47)	0.93 (0.26)
Proportion of learning objectives rated as important or somewhat important, averaged across ACTA users	0.95 (0.43)	0.83 (0.38)
Proportion of student manual modifications rated as accurate, averaged across ACTA users	0.89 (0.31)	0.65 (0.48)
Proportion of learning objectives rated as accurate, averaged across ACTA users	0.92 (0.24)	0.54 (0.50)

Standard deviations are included in parentheses.

based knowledge, as opposed to knowledge that is easily obtained in a classroom setting. The information that newly-trained personnel possess is more likely to have been acquired in a classroom setting rather than through lived experiences.

The findings offer strong support that the ACTA tools allowed students to elicit important and relevant domain information. In the firefighter study, the percentage of content of the cognitive demands tables that was judged to be information that only highly-experienced personnel would know averaged 95%. In the EW domain, the same assessment yielded an average of 90% across the ACTA group. In response to questions regarding information relevance, a substantial percentage of information in the cognitive demands tables ($M = 73\%$) was rated as relevant for a Fireground Commander. Similar results were found in the EW domain where a mean of 87% of the information in the cognitive demands tables was rated as relevant to an EW supervisor.

The third validity question focused on the importance and accuracy of the instructional materials generated by ACTA users. The measures included ratings of importance and accuracy by domain experts. The 3-point importance ratings were collapsed into a dichotomy, with 'important' or 'somewhat important' ratings combined into a single importance indicator. Accuracy had been assessed as a dichotomy (accurate versus not). Findings indicate that in both domains, content of instructional materials generated by ACTA was viewed as important domain information for novices to learn. In the firefighting domain, a mean of 70% of the instructional material modifications generated, and 95% of the learning objectives generated by each student were rated as important. In the EW domain, these means were 93% and 83%, respectively (table 7).

Accuracy evaluations were also high, particularly for the firefighting data. In the firefighting domain a mean 89% of the modifications to the student manuals were rated as accurate and 92% of the learning objectives were rated as accurate. In the EW domains, these means were 65% (modifications to the student manual) and 54% (learning objectives). The authors suspect that the lower accuracy ratings in the EW domain were due to the more technical nature of the domain. The environment in which an EW operator/supervisor works was so foreign to the graduate students that understanding and using the terminology and acronyms that EW operators/ supervisors use to describe the equipment and environment presented additional difficulty. There were a number of cases in which the electronic warfare SMEs rating the data indicated that they knew what the student must have meant, but that the wording used was incorrect.

3.3.3. *Reliability*:   There is no well-established metric or method for assessing the reliability of cognitive task analysis tools, and yet the issue is an important one. Briefly, the question is whether individuals using a particular technique are able to generate comparable information, so that the tools can be considered a source of consistent information, given the same (or similar) domain expert assessed at different points in time and/or by different knowledge engineers. This is a much simpler matter when one is dealing with highly-structured interview formats or scale items than when faced with textual knowledge representations. The authors sought to address the issue in several ways. One approach was to examine whether ACTA users consistently elicited the same types of cognitive information. Therefore, the authors examined the content of the cognitive demands generated by the students, to see whether they had generated similar information.

One set of analyses examined whether ACTA users had generated similar types or categories of cognitive information. This analysis utilized a coding scheme based on Rasmussen *et al.*'s (1994) model of decision making. Each item in each cognitive demands table was coded. In the firefighting domain, every cognitive demands table (100%) generated by the ACTA group contained information that had to do with situation analysis and planning. All but one of the cognitive demands tables (80%) contained data on information collection. Given that students were instructed to focus on the subtask of 'size-up', which consists of gathering relevant information in order to accurately assess the situation and develop a plan of action, it was concluded that students in this study were able to consistently elicit relevant cognitive information using the ACTA techniques.

The same analysis was carried out for the EW study. All ACTA users generated cognitive demands that included information about situation analysis and all but one collected data in the information collection category. The signal threat analysis task consists primarily of gathering the necessary information to maintain an accurate, current assessment of the situation at all times. Again the data indicate that the students consistently elicited relevant cognitive information using the ACTA tools. Data across the two domains suggest that students were able to consistently elicit comparable cognitive information using the ACTA techniques.

A second coding scheme, specific to the firefighting data, also indicated that students consistently elicited similar information. The Fireground Commander task was divided into three subtasks: size-up; strategy and tactics; and management. All the ACTA users obtained information in each of these categories. The bulk of the information gathered focused on the size-up task (62%), which is where the students were asked to focus their interviews. Thus it is concluded that, using ACTA, people were consistently able to get important cognitive information for the entire Fireground Commander task, with an emphasis on the size-up task.

The authors also attempted to assess the degree of overlap of specific items across the cognitive demands tables generated by ACTA users. This proved to be extremely difficult, because users had not been constrained in level of detail, phrasing, or specificity. One student might list as a cognitive demand 'look of the smoke' while another noted 'color and movement of smoke'. The levels of inference required by raters to judge the degree to which any two cognitive demands table items matched were similar or were different became unacceptable, and the analysis was abandoned.

However, an informal examination of the cognitive demands tables suggests that the graduate students did not, in most cases, generate identical cognitive demands. This is not surprising given the design of this study. In order to reduce intra-subject variability, the authors excluded from the study graduate students who had any experience in the domain for which they would be conducting the cognitive task analysis; this meant that all of the students were working at a disadvantage in conducting the CTA. When the ACTA tools are described to professional audiences, it is recommended that time is spent becoming familiar with a domain before interviews are conducted. In this case, the students were given a brief overview of the domain and the task that they would be studying. The limited time we had with the graduate students did not allow for the recommended level of familiarization with the domain. A second reason why the graduate students did not generate the same cognitive demands is that each student was exposed to only two interviews with SMEs. If SME availability had allowed each student access to three to five experts as is generally recommended, the students would have been more likely to have heard similar things in the interviews.

3.3.4. *Group differences*:   One of the drawbacks of the evaluation study design was the limited sample size. The intensive workshop preparation necessary to train subjects in ACTA methods, the extensive coding and data transformation effort necessary to provide empirical evaluation data, and the limited number of available SMEs, made large samples simply beyond the time or resources available. Obviously, with the small group sizes available, the effects associated with membership in the ACTA group were going to have to be very strong to be discernable as statistically significant.

Nonetheless, the authors were surprised to find so few differences between the ACTA group and the Unstructured interview group in the data. In addition to the small sample size, large intra-group differences were found that appear also to account for the lack of statistically significant results. Although an attempt was made to match the groups for age, gender, and education level, considerable individual differences .were found in the students' comfort level and ease in conducting interviews (as observed by the investigators). This resulted in large standard deviations on nearly all the comparative measures, making those findings that were statistically significant difficult to interpret. For example, in rating the evaluation study experience, graduate students in the ACTA group for both the firefighter and the EW study agreed more strongly with the statement, 'I felt confident in my ability to lead an interview' ($M = 4.18$, $SD = 0.60$) than the graduate students who conducted unstructured interviews ($M = 3.25$, $SD = 0.87$), $U = 28.5$, $p = 0.02$. A Mann-Whitney U-test, which is free of variance assumptions, was used instead of a t-test because there was no variance in the responses from the ACTA students in the electronic warfare domain. In the firefighter study, the ACTA group agreed more strongly with the statement, 'I felt I had sufficient information to revise the course materials' ($M = 4.00$, $SD = 0.89$) than the Unstructured group ($M = 2.67$, $SD = 1.03$), $t(10) = 2.39$, $p = 0.06$. These statistical analyses indicate that students trained to use ACTA felt more confident in conducting the interviews and were more confident that they had gathered sufficient information to revise the course materials than the Unstructured group. However, in looking at the large standard deviations, it becomes clear that some students in each group were confident, whereas others were not.

In other cases, the means indicate very little difference between the groups, but the standard deviations indicate considerable variance within the groups. For example, the means for the two groups are nearly identical in response to the question, 'Given the information, I found the revision of course materials to be straightforward'. However, the large standard deviations indicate that some people in each group found the revision of the course materials straightforward, but others did not (table 8). Given the small sample sizes used in this study, it is clear that these group difference comparisons are not very robust.

One potentially confounding factor in the design of the study was that during the introductory workshop, the Unstructured group was exposed to a lecture on cognitive elements, cognitive task analysis, and how to fill out a cognitive demands table before conducting interviews. Examination of the raw data suggests that some of the students in the Unstructured group may have used the cognitive demands table to structure their interviews, thus reducing the gap between the ACTA group

Table 8. Mean responses to Question 10 on the interviewer questionnaire.

	ACTA group		Unstructured group	
	Firefighting domain	Electronic warfare domain	Firefighting domain	Electronic warfare domain
Given the information, I found the revision of course materials straightforward	3.67 (0.82)	3.83 (1.47)	3.00 (1.23)	3.33 (1.03)

Responses refer to a 5-point rating scale, where, 1 = strongly disagree, 5 = strongly agree. Standard deviations are in parentheses.

and the Unstructured group. The implication for ACTA is that the cognitive demands table is a valuable tool for framing the kinds of information that the interviewer intends to elicit from the SME.

Although the authors considered using a control group that would receive no introductory workshop on cognition, this was found to be impractical given that they wanted to compare both the amount of cognitive information elicited in the interviews and the quality of the training materials produced. In order to make these comparisons, it was necessary to provide training in how to create a cognitive demands table and how to produce instructional materials to all of the participants in the study.

The high quality ratings (i.e. SME ratings of importance and accuracy) received by both interview groups indicate that an exposure to the concepts underlying cognitive task analysis and a description of how cognitive task analytic data can be applied to instructional materials, may play a large role in learning to conduct cognitive task analyses. Working only with this foundational material, in the absence of exposure to actual methodologies, some students in the Unstructured group were able to gather accurate, relevant cognitive information and develop useful instructional materials.

4. Discussion

The findings presented here indicate that, after a 6-h workshop introducing the ACTA techniques, graduate students were able to conduct interviews with SMEs and elicit important, accurate cognitive information that was easily translated into instructional materials. Furthermore, subjective usability data indicate that the graduate students found the techniques to be easy to use, flexible, and to provide clear output. The authors' belief is that professional instructional designers and systems designers will do even better than the graduate students, given that they will have more concrete goals for use of the cognitive information and more experience in generating applications.

Although an attempt has been made here to establish the reliability and validity of the ACTA methods, the authors are aware that no well-established metrics exist. The need to test CTA methods in real-world settings with real-world tasks greatly reduces the level of control that one has over the many sources of variability. Factors that are difficult to control include the fact that some people seem to be more predisposed to be good interviewers than others. In addition, some SMEs are more articulate and easier to focus than others. Given the variability among humans in both the interviewer and the SME roles, it will be important to answer such questions as: Does an SME report the same examples and the same details when asked the same question later in time? Do the CTA techniques elicit the same types of information when used by different interviewers? Do independent practitioners generate the same materials based on CTA interviews? Further work is need to establish meaningful metrics to assess the reliability and validity of CTA tools.

An important point to make here is that although the ACTA methods have been shown to elicit important, accurate cognitive information, the authors have yet to try to assess what is lost using these streamlined techniques. It is believed that a trade-off exists: the more streamlined and proceduralized CTA techniques become, the less powerful they are. It appears that the ACTA techniques gather less comprehensive information than more systematic techniques such as the PARI method (Hall *et al.* 1994) and Gordon and Gill's (1992) conceptual graph analysis, and that the information gathered is more superficial than that gathered using the critical decision method (Klein *et al.* 1989) or Rasmussen's (1986) cognitive analysis. In spite of the limitations of streamlined CTA procedures, the ACTA techniques provided graduate students with sufficient tools to identify key cognitive elements and develop useful training materials. Until better metrics exist, however, it will be difficult to objectively assess what is lost and what is gained via different techniques.

It is also important to point out the impact of a failure to investigate and incorporate cognitive issues in complex tasks requiring high degrees of cognitive skill. Despite the promise of automated and intelligent systems, the human decision-maker will always play a role in systems where uncertainty and ambiguity exist. The consequences of not training operators to acquire the cognitive skills required, or not designing systems to support human problem-solving and decision-making can be dire, as illustrated by disasters such as Three Mile Island, the USS Vincennes, and Kings Cross Station, to name but a few (Reason 1990).

The ACTA methodology was originally taught in a workshop format, which allowed for the workshop presenters to tailor the methods to the audience and add personal anecdotes to the instruction. This also meant that no two ACTA workshops were the same. This made research into the reliability of the methods even more difficult to evaluate. However, more recently the ACTA workshop has been produced on a compact disk-based, multimedia training tool (Militello *et al.* 1997). This tool provides the learner with the reasons for undertaking a cognitive task analysis, an introduction to cognition and expertise, and a tutorial on the three ACTA methods. This tool provides an opportunity to conduct a more controlled comparison of the effectiveness and reliability of the ACTA methodology. Future research issues include looking at the reliability and validity of the knowledge elicitation and representations, as well as of the end-product training and systems interventions.

Future research using ACTA should also explore the following areas: incorporating methods for analysing tasks that rely implicitly on multi-operator interactions and teamwork; improving the incorporation of ACTA into more traditional and systematic task analytic techniques; improving the knowledge representations so that they are more easily translated into training and system design interventions; and, improving the training methods, and system and interface design interventions themselves, so that cognitive strengths and cognitive vulnerabilities can be better supported.

Acknowledgements

This research was carried out under contract number N66001-94-C-7034 from the Navy Personnel Research and Development Center, San Diego, CA, USA.

References

BELL, J. and HARDIMAN, R. J. 1989, The third role—the naturalistic knowledge engineer, in D. Diaper (ed.), *Knowledge Elicitation: Principles, Techniques, and Applications* (New York: Wiley), 49–85.

CHI, M. T. H., FELTOVICH, P. J. and GLASER, R. 1981, Categorization and representation of physics problems by experts and novices, *Cognitive Science*, **5**, 121–152.

CLARKE, B. 1987, Knowledge acquisition for real-time knowledge-based systems, *Proceedings of the First European Workshop on Knowledge Acquisition For Knowledge Based Systems* (United Kingdom: Reading University), September.

COOKE, N. J. 1994, Varieties of knowledge elicitation techniques, *International Journal of Human-Computer Studies*, **41**, 801–849.

CORDINGLEY, E. S. 1989, Knowledge elicitation techniques for knowledge-based systems, in D. Diaper (ed.), *Knowledge Elicitation: Principles, Techniques, and Applications* (New York: Wiley), 89–175.

CRANDALL, B. and CALDERWOOD, R. 1989, Clinical assessment skills of experienced neonatal intensive care nurses. Final report, Klein Associates Inc., OH. Prepared under contract 1 R43 NR0191101 for The National Center for Nursing, NIH.

CRANDALL, B. and GAMBLIAN, V. 1991, Guide to early sepsis assessment in the NICU. Instruction manual prepared for the Ohio Department of Development under the Ohio SBIR Bridge Grant program by Klein Associates Inc., OH.

CRANDALL, B. and GETCHELL-REITER, K. 1993, Critical decision method: A technique for eliciting concrete assessment indicators from the 'intuition' of NICU nurses, *Advances in Nursing Sciences*, **16**, 42–51.

DIEDERICH, J., RUHMANN, I. and MAY, M. 1987, KRITON: A knowledge acquisition tool for expert systems, *International Journal of Man-Machine Studies*, **25**, 29–40.

DREYFUS, H. L. 1972, *What Computers Can't Do: A Critique of Artificial Reason* (New York: Harper & Row).

DREYFUS, H. L. and DREYFUS, S. E. 1986, *Mind over Machine: The Power of Human Intuitive Expertise in the Era of the Computer* (New York: The Free Press).

DRURY, C. G., PARAMORE, B., VAN COTT, H. P., GREY, S. M. and CORLETT, E. N. 1987, Task analysis, in G. Salvendy (ed.), *Handbook of Human Factors* (New York: Wiley), 370–401.

FLANAGAN, J. C. 1954, The critical incident technique, *Psychological Bulletin*, **51**, 327–358.

GILBRETH, F. B. 1911, *Motion Study* (New York: Van Nostrand).

GILBRETH, F. B. and GILBRETH, L. M. 1919, *Fatigue Study* (New York: Macmillan).

GORDON, H. W. and GILL, R. T. 1992, Knowledge acquisition with question probes and conceptual graph structures, in T. Laurer, E. Peacock and A. Graesser (eds), *Questions and Information Systems* (Hillsdale, NJ: Lawrence Erlbaum), 29–46.

GROVER, M. D. 1983, A pragmatic knowledge acquisition methodology, *Proceedings of the 8th International Joint Conference on Artificial Intelligence*, Karlsruhe, Germany, August, 436–438.

HALL, E. M., GOTT, S. P. and POKORNY, R. A. 1994, A procedural guide to cognitive task analysis: the PARI method, unpublished manuscript, Brooks AFB, TX.

HOFFMAN, R. R. 1987, The problem of extracting the knowledge of experts from the perspective of experimental psychology, *AI Magazine*, **8**, 53–67.

HOFFMAN, R. R. 1992, *The Psychology of Expertise: Cognitive Research and Empirical AI* (New York: Springer-Verlag).

HOFFMAN, R. R., CRANDALL, B. E. and SHADBOLT, N. R. 1998, A case study in cognitive task analysis methodology: The critical decision method for the elicitation of expert knowledge, *Human Factors* (In press).

HOFFMAN, R. R., SHADBOLT, N. R., BURTON, A. M. and KLEIN, G. 1995, Eliciting knowledge from experts: A methodological analysis, *Organizational Behavior and Human Decision Processes*, **62**, 129–158.

HOWARD, G. S. 1994, Why do people say nasty things about self-reports? *Journal of Organizational Behavior*, **15**, 399–404.

HOWELL, W. C. and COOKE, N. J. 1989, Training the human information processor: A look at cognitive models, in I.L. Goldstein (ed.), *Training and Development in Work Organizations: Frontiers of Industrial and Organizational Psychology* (San Francisco: Jossey-Bass), 121–182.

KAEMPF, G. L., KLEIN, G., THORDSEN, M. L. and WOLF, S. 1996, Decision making in complex command-and-control environments, *Human Factors Special Issue*, **38**, 220–231.

KIRWAN, B. and AINSWORTH, L. K. 1992, *A Guide to Task Analysis* (London: Taylor & Francis).

KLEIN, G. 1993, Naturalistic decision making—implications for design. SOAR 93-01 (Contract no. DLA900-88-0393) CSERIAC, OH. State-of-the-Art Report 93-01; Dayton, OH: Crew System Ergonomics Information Analysis Center.

KLEIN, G. A. and HOFFMAN, R. 1993, Seeing the invisible: perceptual/cognitive aspects of expertise, in M. Rabinowitz (ed.), *Cognitive Science Foundations of Instruction* (Hillsdale, NJ: Lawrence Erlbaum), 203–226.

KLEIN, G. A., CALDERWOOD, R. and MACGREGOR, D. 1989, Critical decision method for eliciting knowledge, *IEEE Transactions on Systems, Man, and Cybernetics*, **19**, 462–472.

KLINGER, D. and GOMES, M. 1993, A cognitive systems engineering application for interface design, *Proceedings of the 37th Annual Meeting of the Human Factors Society* (Santa Monica, CA: Human Factors Society).

MEISTER, D. 1985, The two worlds of human factors, in R. E. Eberts and C. G. Eberts (eds), *Trends in Ergonomics/Human Factors II*, (New York: North-Holland), 3–11.

MILITELLO, L. G. and LIM, L. 1995, Early assessment of NEC in premature infants, *Journal of Perinatal and Neonatal Nursing*, **9**, 1–11.

MILITELLO, L. G., HUTTON, R. J. B. and MILLER, T. 1997, [Computer software] *Applied Cognitive Task Analysis* (Fairborn, OH: Klein Associates Inc).

NISBETT, R. E. and WILSON, T. D. 1977, Telling more than we can know: Verbal reports on mental processes, *Psychological Review*, **84**, 231–259.

PEDHAZUR, E. J. and SCHMELKIN, L. P. 1991, *Measurement, Design, and Analysis: An Integrated Approach* (Hillsdale, NJ: Lawrence Erlbaum).

RASMUSSEN, J. 1986, *Information Processing and Human-machine Interaction: An Approach to Cognitive Engineering* (New York: North-Holland).

RASMUSSEN, J., PEJTERSEN, A. M. and GOODSTEIN, L. P. 1994, *Cognitive Systems Engineering* (New York: Wiley).

REASON, J. 1990, *Human Error* (New York: Cambridge University Press).

ROTH, E. M., WOODS, D. D. and POPLE, H. E., Jr. 1992, Cognitive simulation as a tool for cognitive task analysis, *Ergonomics*, **35**, 1163–1198.

ROUSE, W. 1984, Models of natural intelligence for fault diagnosis tasks: Implications for training and aiding of maintenance personnel, in Artificial Intelligence in Maintenance: Proceedings of the Joint Services Workshop, report no. AFHRL-TR-84-25, Air Force Human Resources Laboratory, Colorado, 193–212.

ROUSE, W. B. and MORRIS, N. M. 1986, On looking into the black box: Prospects and limits on the search for mental models, *Psychological Bulletin*, **100**, 349–363.

SEAMSTER, T. L., REDDING, R. E., CANNON, J. R., RYDER, J. M. and PURCELL, J. A. 1993, Cognitive task analysis of expertise in air traffic control, *The International Journal of Aviation Psychology*, **3**, 257–283.

SEAMSTER, T. L., REDDING, R. E. and KAEMPF, G. L. 1997, *Applied Cognitive Task Analysis in Aviation* (Aldershot: Avebury Aviation).

SHANTEAU, J. 1985, Psychological characteristics of expert decision makers, *Applied Experimental Psychology Series*, **85** (Kansas State University, Kansas).

SPECTOR, P. E. 1994, Using self-report questionnaires in OB research: A comment on the use of a controversial method, *Journal of Organizational Behavior*, **15**, 385–392.

TAYLOR, F. W. 1911, *The Principles of Scientific Management* (New York: Harper).

TAYNOR, J., CRANDALL, B. and WIGGINS, S. 1987, The reliability of the critical decision method, KATR-863(B)-87-07F. Prepared under contract MDA903-86-C-0170 for the US Army Research Institute Field Unit, Leavenworth, KS, by Klein Associates Inc., OH.

THORDSEN, M. L., MILITELLO, L. G. and KLEIN, G. A. 1992, Cognitive task analysis of critical team decision making during multiship engagements Prepared under subcontract RI-82264X for Williams Air Force Base, AZ, by Klein Associates Inc., OH.

CHAPTER 7

Enhancing the usability of a task analysis method: a notation and environment for requirements specification

Thomas C. Ormerod†*, Juliet Richardson†§ and Andrew Shepherd‡

†Department of Psychology, Lancaster University, Lancaster LA1 4YF, UK

‡Department of Human Sciences, Loughborough University, Loughborough, UK

Keywords: Hierarchical task analysis; Sub-goal templates (SGTs); Information requirement specification; Design tools and methods; Usability; Interface design; Process control.

This paper presents a notation and computer-based tool for the Sub-Goal Template (SGT) method, a task-analytic approach to specifying information requirements. In particular, it focuses upon the ergonomic redesign of a notation used in the SGT method for specifying contingent sequences in operators' tasks. Two experiments are reported in which two notations for redescribing sequences that involve making a choice between task alternatives are compared: a disjunctive form (*Either...or...*) used in the original SGT scheme, and a conditional form with a redundant negative (*If...then...if not...*), which has been promoted as a notation for capturing procedural sequences in computer programming. In experiment 1, performance with the conditional notation was better than with the disjunctive notation for redescribing simple and moderately complex task plans, although there was no difference for the most complex task plans. In experiment 2, a computer environment for specifying task plans using the SGT method was compared with a paper-based equivalent. In general, the computer environment was found to be easier to use and led to greater accuracy in plan redescription, although it was slower perhaps as a result of the constraints that it placed on participants. This paper argues for the necessity of ergonomic design in developing notations and tools to support task analysis methods.

1. Introduction

A number of methodologies have been developed for specifying user requirements (Payne and Green 1986, Card *et al.* 1983, Lim and Long 1994). While usability methods have a worthy aim, they are often curiously insensitive to the usability issues that they themselves generate. This is particularly true with regard to task analysis. Almost all user-oriented methodologies embody some aspect of task analysis, whether explicitly or implicitly, and some (e.g. the task-action grammar method of Payne and Green 1986) provide notations for describing the *products* of task analysis. Yet few, if any, of these methodologies provide a notation that is developed to support the *process* of task analysis. Notations that are intended to enhance the specification of requirements are often developed without regard to their

---

* Author for correspondence.

§* *Current address*: School of Psychology, University of Wales College of Cardiff, PO Box 901, Cardiff CF1 3YG, Wales.

own usability, despite the fact that research has shown the potential value of evaluating notations (Greene *et al.* 1990).

An important part of a user's requirements is their 'information requirements'; that is, the information that must be displayed in order for a user to operate a system effectively, together with details on how to organize that information sequentially. Information requirements are often sidelined in the design process: many methodologies support only the specification of functional requirements (Harker *et al.* 1990). This paper describes how one task-analytic method, the Sub Goal Template (SGT) method of Shepherd (1993; see also Shepherd and Ormerod 1992) was redesigned to enhance its usability. The SGT method adopts Hierarchical Task Analysis (HTA; Annett *et al.* 1971) as the basis for information requirements specification. Under the SGT method, the information requirements for each of the tasks specified in an HTA are determined through the application of task templates, to which the user need only supply task parameters to complete stereotypical information requirements for each task.

The structure of the paper is as follows: first, the authors outline the SGT method, in particular introducing a design cycle that outlines the use of the SGT method as an integral part of HTA. Next, the authors describe how a shortcoming in the notation chosen for specifying sequence information in the original SGT scheme necessitated its redesign. The new notation is evaluated empirically in experiment 1, where it is shown to overcome many of the problems of the original sequencing notation. Finally, the development of a computer tool for the SGT method is described and this is evaluated against a paper-based equivalent in experiment 2.

2. The SGT method

Sub Goal templates (SGTs) are a set of standard task elements that capture most, if not all, of the tasks that control operators encounter in any process plant (see table 1 for a description). The purpose of the scheme is four-fold (see Shepherd (1993) for further discussion). First, it provides a vocabulary for redescribing the outcomes of task analysis, which facilitates the process of task identification and introduces a degree of systematicity and rigour into task labelling and organization. Second, the application of standard task templates during the process of HTA provides a clear

Table 1. SGT task and sequencing elements (from Shepherd and Ormerod 1992).

Code	Label	Task types/Syntax
Task elements		
A	Action	Activate (A1), Adjust (A2), De-activate (A3)
C	Communication	Read (C1), Record (C2), Wait for information (C3), Receive information (C4), Give information (C5), Remember (C6), Retrieve (C7)
M	Monitoring	Monitor to detect deviance (M1), Monitor rate of change (M2), Inspect plant and equipment (M3)
D	Decision	Diagnose process problems (D1), Adjust plan (D2), Locate contaminant (D3), Judge adjustment (D4)
Sequence elements		
S1	Fixed	*S1 Then* X
S2	Contingent	*S2 Either* Z *Then* X *Or* not Z *Then* Y
S3	Parallel	*S3 Then do together* X *and* Y
S4	Free	*S4 In any order* X *and* Y

stopping point at which task decomposition can cease. This is particularly important in the light of critics of task analysis (e.g. Benyon 1992), who argue that the outcomes of task analysis risk being device-dependent, hence limiting the search for creative design solutions. SGTs are device-independent, and do not carry with them any implied or explicit implementation. Third, each task element has a unique set of information needs (for examples, see Lansdale and Ormerod 1994). Labelling tasks with SGT elements entails the automatic specification of information requirements for successful task completion without constraining the design options that deliver the task's information needs. In this way, the SGT scheme provides a full requirements specification for the interface designer as an automatic outcome of HTA. Fourth, SGT task elements are accompanied by a set of sequencing elements (table 1), that enable the description of task orders, dependencies and contingencies in a clear and precise fashion.

The SGT sequencing elements are the focus of the empirical studies reported in this paper. The importance of capturing the orders, dependencies and contingencies that exist in operator tasks was highlighted by Shepherd (1986) in his use of plans for HTA. Plans describe the logical order in which the sub-tasks that constitute a task must be undertaken depending upon the conditions of task performance. The SGT sequencing elements provide a notation for describing task plans in terms of four different relationships between tasks: fixed, contingent, parallel and free sequences. The function of these elements is to specify the way in which information should be arranged for presentation to the operator. For example, if two operations are carried out in parallel then the information needs of both must be displayed together.

Figure 1 shows an example of a paint mixer plant (from Shepherd (1990): this example is also used in the experiments reported in this paper). A part-HTA for the mixer plant is shown in figure 2. Operating this plant has the immediate sub-goals 'Clean out plant', 'Start up plant', 'Monitor and maintain plant parameters' and 'Carry out shutdown'. Each of these sub-goals has associated tasks that can be further redescribed. Redescription ceases at the level at which sub-goals correspond to tasks that have 'no problem-solving or control structure component' (Benyon 1992:105); usually these are individual operations such as reading fluid levels or closing valves. HTA can be used to gather information regarding users' tasks and the constraints acting on these. For example, in the mixer plant the process can only be fed forward after the equipment has been prepared. This constraint is described in Plan 1.2 in figure 2.

The redescription using SGT sequencing elements of part of the mixer plant HTA is illustrated in figure 3. This shows what the operator's tasks are, the orders in which they must be carried out and the dependencies that exist between them.

The SGT scheme was originally conceived as a notation for redescribing a completed HTA so that information requirements for all tasks could be collated. Thus, the original method of applying the scheme (and the task given to participants in the experiments reported in this paper) was to translate task plans from completed HTAs into the SGT notation. However, there are a number of advantages to making the application of SGTs an integral part of HTA. For example, it makes requirements specification a single rather than a two-stage process, and it provides a structure around which the process of HTA can be carried out in a rigorous fashion, ensuring that analysis does not cease too early before sufficient task information is collected, or too late so that excessive and/or irrelevant information is collected. Hence a distinction can be made between the SGT 'scheme', which describes the notation, and the SGT 'method', which describes the use of the scheme during an ongoing HTA.

Figure 4 shows the design cycle that forms the basis of the SGT method. In essence, the SGT method works as follows: the analyst searches for the highest level task. An attempt is then made to assign one of the SGT task elements to it. If this fails, then the task must be decomposed into sub-tasks. At this stage, a plan must be constructed using the SGT sequencing notation that shows how sub-tasks are ordered for successful task completion in each possible contingency. The process then cycles, with an attempt to assign an SGT to each sub-task or further decomposition and data collection, where necessary. When an SGT is assigned successfully to a task, decomposition ceases at this level, and a set of stereotypical

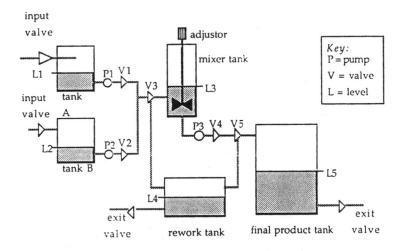

In this plant two liquids, A and B, are blended together in a specific ratio to form a product. Liquids A and B are stored in feed tanks which are topped up from a store farm as needed. The liquids are mixed together in the mixer tank by a stirrer and the resulting product is stored in the final product tank. Any mixture that is out of specification is diverted to the rework tank where it is later collected and reprocessed. Flow through the pipes is driven by pumps 1, 2 and 3 and is controlled by needle-valves 1, 2 and 4. For example, pump 1 pumps liquid A from its feed tank towards the mixer tank and the rate of this flow is controlled by valve 1. Valves 3 and 5 are diverter valves and can be used to divert liquids to the rework tank. The following technical and operational constraints control the operation of the Mixer Plant:-

- The pumps and the stirrer must be started up under load.
- Liquids A and B must be blended in the ratio 5:1 to produce 500 litres of product.
- Tank A and tank B can be filled to 800 litres and 200 litres respectively.
- The ratio and blend of the product is critical. Therefore if there is any possibility of error it should be switched to the rework tank.
- In order to maximise efficiency the batch should be completed as quickly as possible and the minimum possible amount of product should be sent to the rework tank.

Figure 1. Illustration of the mixer plant (after Shepherd 1990:5).

information requirements are assigned to the task. The computer tool that is reported in this paper adopts this design cycle as its underlying process model.

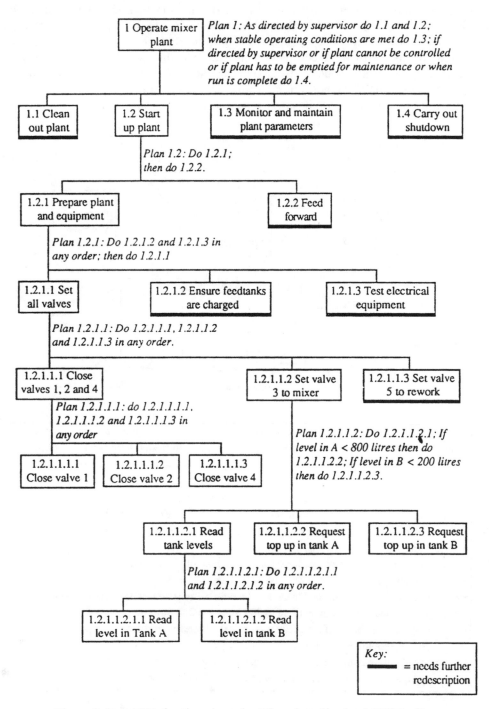

Figure 2. Part-HTA for the mixer plant (based on Shepherd 1990:7-8).

The SGT scheme has been used successfully in a number of projects, for process control applications (Shepherd and Ormerod 1992) and for computer telephony dialogue design (Webb 1996). The scheme has also been independently shown to perform more successfully than an in-house method for information requirements specification (Schaftel 1994). Furthermore, the authors have recently argued that the SGT scheme overcomes the problems raised by Benyon (1992) in using task analysis as a systems design methodology (Richardson *et al.* 1998). In particular, the authors have argued that the SGT scheme presents a data-centred method for requirements specification, yet it avoids device-dependent design through the provision of templates that act as stopping rules in task decomposition.

Set valve 3 to mixer (1.2.1.1.2 in Figure 2).

S3   **Then do together**

     1.2.1.1.2.1

and

     S2   **Either** level A < 800 litres

     Then

         1.2.1.1.2.2

     Or level A > 800 litres

     Then

         continue

and

     S2   **Either** level B < 200 litres

     Then

         1.2.1.1.2.3

     Or level B > 200 litres

     Then

         continue

Figure 3. Redescription of a section of the HTA for mixer plant operation using sequencing elements.

3.   Re-design of the SGT sequencing notation

Despite the promise of the SGT scheme, preliminary usability evaluations indicated certain problems in its application as a method for specifying user information requirements. Shepherd and Ormerod (1992) found that experienced plant operators and interface designers were often unable to redescribe operating plans using appropriate SGT sequencing elements when plans contained an implicit conditional statement such as 'Start in-tank pump when pressure reaches 8 bar then adjust bleed valve to vaporisers to maintain 8 bar'. This problem concerns the 'contingent' sequencing element S2 (table 1), which is used when a choice of possible actions depends upon the outcome of a previous action or event. In the original SGT scheme, this relationship was expressed through the use of a disjunctive (*Either...or...*) statement. For example, the operational sequence 'If the pump alarm flashes then close the valve' would be rewritten as:

S2 Either
  • pump alarm flashes
Then
  • close the valve
Or
  • pump alarm does not flash
Then
  • continue

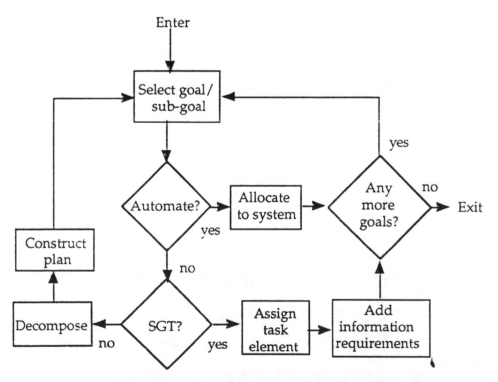

Figure 4. The design cycle underlying the SGT method.

The disjunctive construction was chosen for the SGT scheme because it makes clear the need to specify explicitly what happens in the negative case (e.g. when the alarm does not go off). The difficulty of using a disjunctive construction was confirmed by Richardson and Ormerod (1997), who found that participants encountered difficulties when asked to rewrite a conditional expression as a disjunctive and vice versa. If the SGT scheme is to be used successfully as a requirements specification method, then it must have a notation that can be used easily and accurately. In order to achieve this, the S2 element required redesign to remove the disjunctive form whilst retaining the ability to express the negative case.

Contingent sequences are commonly found in computer programs, so it seems appropriate to consider notations that are available in the programming domain in the search for an alternative contingent sequence notation. Sime *et al.* (1977c) compared three types of 'micro-language', each containing a different notation for expressing conditional statements: a JUMP notation of the form *if..then goto. . .*; a NEST-BE notation of the form *if..then begin..end..else begin ..end*; and a NEST-INE notation of the form *if..then..if not..then..end*. There are several differences between these languages including their spatial layout and the number of stopping statements that they include to indicate the scope of the conditionals. However, the main difference emphasized by Sime *et al.* (1977c) was that the NEST-BE notation implicitly states the negative contingency using the word 'else', whilst the NEST-INE language contains an explicit, though computationally redundant, negative. In comparison, the JUMP language contains no statement of the negative contingency. Sime *et al.* (1977c) report that novices made fewer errors in writing programs using the NEST languages. They also found that participants corrected errors fastest with the NEST-INE language (see also Green (1977) who found similar advantages for the NEST-INE with professional programmers). Sime *et al.* (1977c) argue that programming involves translating 'taxon' information into sequences. Taxon information concerns the conditions under which actions can occur. Debugging involves the reverse of program writing, the translation of sequences into taxon information. NEST languages make sequence information clear for program writing, and redundant negatives in the NEST-INE language make taxon information clear for program debugging. It could be argued that differences in the interpretability of the different NEST languages might be due to the presence of extra stopping statements in the NEST-BE language, rather than the presence of redundant negatives in the NEST-INE language. However, intuitively, one would predict that the extra stopping statements would make the NEST-BE language easier to interpret, which is the opposite of what Sime *et al.* (1977c) actually observed. Instead, it would seem likely that participants might interpret the '*if not..*' in the NEST-INE language as an implicit stopping statement for the positive case, thus making the two languages essentially equal in this respect.

The tasks of programming and plan redescription in the SGT scheme are analogous in one important respect: both involve the translation of information about the conditions under which actions or outcomes arise into procedural sequences. Thus, the S2 element used in the SGT scheme was redesigned with a NEST-INE conditional notation, which retains the redundant negative while removing the awkward disjunctive construction. The new S2 element takes the following form (where X and Y are tasks and Z is an operational condition):

S2 If   Z
    Then   X
  If not   Z
    Then   Y

4. Experiment 1: Evaluation of the redesigned notation

The main aim of this evaluation was to compare the use of the two versions of the S2 notation for redescribing contingent sequences. Participants were split into two groups: one group used the original disjunctive notation to redescribe sequences of operational activities while the other group used the redesigned conditional notation. The two groups received the same instructions and materials: the only difference between them was in the notation of the S2 element that they used.

A pilot study showed no difference between notations when participants were required to redescribe simple sequences of actions, but problems arose when they were presented with more structurally complex sequences. Thus, in this experiment participants were presented with sequences of varying structural complexity. Complexity was varied by manipulating the depth to which tasks were embedded within other task sequences, in a way similar to the nesting of procedural loops in computer programs. Each of the tasks to be redescribed by participants involved an operational sequence based on the mixer plant (figure 1). Unfamiliar task descriptions were used to ensure that participants did not differ in terms of domain-specific knowledge about the sequences. For example, with a familiar task such as operating a microwave oven, some participants may set the timer first while others may set the power level first.

4.1. *Method*

4.1.1. *Participants*:   A total of 26 undergraduate students from the Psychology Department at Lancaster University were paid £5 each to participate in this study. None of the participants had experience of programming, process control or engineering design. The use of naïve participants ensures that they cannot simply bring background knowledge to bear in carrying out the tasks. Thus, the use of the SGT scheme in this experiment is not helped or hindered by participants' prior experience, either with process control plants, or with using similar notations. The use of naïve participants is further justified by the demonstration by Green (1977) that the notational effects found with naïve participants by Sime *et al.* (1977c) generalized to professional programmers.

4.1.2. *Materials and design*:   Two factors were manipulated in this experiment—the between-participants factor of Notation with two levels (disjunctive versus conditional) and the within-participants factor of Complexity with three levels (one, two or three embedded elements in the task sequence, as illustrated in table 2). Each task redescription required the use of three of the SGT sequence elements. Each participant received six task redescription problems, two at each level of complexity.

4.1.3. *Procedure*:   Each participant was presented with a glossary containing the sequencing elements, instructions, an illustration of the mixer plant and an example SGT redescription of one mixer plant task. Participants also received a practice problem in which they produced an SGT redescription for the task of boiling an egg.

Participants were allowed to refer freely to the instructions and glossary throughout the experiment. However, they were not allowed to refer back to either the worked example or the practice translation during the experiment in order to prevent them from simply copying these answers. In addition, the purpose of the illustration was only to provide a background for the plan redescriptions; therefore, in order to prevent the illustration being used to interpret the sequences, it was removed as soon as each participant started to attempt the redescriptions.

The six problems were presented in a booklet, one per page, in a different order for each participant. Participants wrote their redescriptions in the booklet. Once the participant had produced a redescription the experimenter informed him/her

Table 2. Examples of the experimental problems and their SGT redescriptions.

Description		Redescriptions		
		Conditional		Disjunctive
1. *embedded task* Start plant production. Then maintain flow through the plant while checking for any problems. Shut down the plant if the supervisor requests it.	S1	Then Start production	S1	Then Start production
	S3	Then do together maintain flow and check for problems	S3	Then do together maintain flow and check for problems
	S2	If receive request then shutdown If not receive request then continue	S2	Either receive request then shutdown Or not receive request then continue
2. *embedded tasks* Shut off the input to the mixer tank and switch the output to the storage tank. Monitor the level of liquid in the mixer tank, when it falls to zero then switch off the output.	S3	In any order shut off input and switch output	S3	In any order shut off input and switch output
	S3	Then do together monitor level and S2 If zero then switch off If not zero then wait	S3	Then do together monitor level and S2 Either zero then switch off Or not zero then wait
3. *embedded tasks* If the alarm sounds, call the supervisor whilst closing all the valves and switching off all the electrical equipment	S2	If alarm sounds then S3 Then do together call supervisor and S4 In any order close valves and switch off If not alarm then continue	S2	If alarm sounds then S3 Then do together call supervisor and S4 In any order close valves and switch off Or not alarm then continue

whether or not it was correct and the time was recorded. If the solution was incorrect then the participant was asked to try and correct his/her errors and the time taken to do this was recorded. Participants continued correcting errors either until a correct answer was produced or until no further possible corrections could be seen by the participant, at which point s/he was shown an example of a correct answer to the problem. Participants worked through the translations at their own pace and were not allowed to refer back to any of their earlier answers once they had been completed. The duration of the experiment was approximately 1 h.

4.2. *Results and discussion*

Participants' redescriptions were judged using two criteria: the attempt had to be free from syntactic errors, and the elements used in the attempt had to be consistent with an appropriate semantic interpretation of the sequence. Table 3 summarizes the mean performance of participants in each Notation group in terms of the following performance measures: number of tasks correctly redescribed, number of redescriptions correct on the first attempt, error lifetimes, time to produce correct solutions, and time taken on trials where an error-free solution was produced at the first attempt. As table 3 indicates, significantly better performance was shown on all measures apart from error lifetimes by participants using the conditional rather than the disjunctive notation.

Participants' errors were analysed in terms of whether they were syntactic or semantic using the same criteria to make this classification as Sime *et al.* (1977c). Syntactic errors are those in which an element is written down incorrectly or used in an incorrect situation. An example of a syntactic error observed in the experiment is:

> *S3 Then do together*
> *switch flow to storage tank*
> *switch off pumps*

In this example the word 'and' has been omitted from the S3 element. While this error might seem minor, in that the word 'and' could be seen to be implied within the text and to therefore be redundant, it should be stressed that notations such as the SGT scheme are designed to be used by many people within a design project. In

Table 3. Means and *t*-test results for measures of accuracy and solution-time (in seconds) in experiment 1.

Measure	Conditional notation		Disjunctive notation			
	Mean	SD	Mean	SD	*t*	Significance (one-tailed)
% Correct solutions	75.0%	19.5%	50.0%	22.6%	2.991	$p < 0.05$
% Correct on first attempt	55.0%	22.8%	36.7%	23.7%	2.032	$p < 0.05$
Time to correct solution	157	65	197	80	2.678	$p < 0.05$
Time to solution correct on first attempt	135	44	170	59	2.860	$p < 0.05$
Error lifetimes	68	41	78	37	1.098	N.S.

N.S. = Not Significant.

other words, one or more people might write a specification using the SGT scheme, which will then be used by another set of people to produce a design. It is therefore important that the notation is used consistently and accurately: what might seem to be implied within a statement by one person, could well be opaque to another. In addition, although individual errors, such as the one illustrated above, might seem to be minor in isolation, they become increasingly important when they occur in the middle of complex embedded plans.

Semantic errors are those in which the answer is syntactically correct but where an incorrect sequence is represented in the redescription. These errors typically involved the use of a wrong element. Participants using the disjunctive notation made significantly more syntactic errors (mean = 6.4 errors) than participants using the conditional notation (mean = 2.3 errors), $t = 3.016$, $p < 0.05$. Table 4 shows the mean number of errors made by participants with each of the sequencing elements. There were no significant differences between conditional and disjunctive groups in the numbers of errors made involving the S1, S3 and S4 elements. The syntax of the S2 element would not be expected to affect the use of these other elements. This finding therefore demonstrates that the performance of the two groups of participants was equivalent, apart from the effects of the manipulation of notation. However, significantly fewer errors involving the S2 element were made by participants using the conditional than the disjunctive notation. Errors involving the S2 element are further classified in table 4 as omission, addition or confusion of words, actions, conditions or elements. Fewer errors of each type were made by participants using the conditional notation.

The number of semantic errors made by participants using the disjunctive notation (mean = 2.14 errors) and the conditional notation (mean = 2.33 errors) did not differ significantly. With both notations, participants knew equally well *when* to use the S2 element: thus, the difference appears to be in knowing *how* to use the notation. Sime *et al.* (1977c) report a similar finding: the same number of semantic errors were made by participants using NEST-BE and NEST-INE languages, but fewer syntactic errors were made by participants using the NEST-INE language. These results indicate the limits of the effects that notations have upon performance. A good notation does not help participants to a better interpretation of the problem; it can, however, help them to express more accurately the interpretation that they have made.

Table 4. Mean numbers of errors involving each SGT sequencing element in experiment 1.

Sequencing element	Conditional notation		Disjunctive notation			
	Mean	SD	Mean	SD	t	Significance
S1 (fixed)	2.00	1.28	3.21	2.08	1.754	N.S.
S2 (contingent)	1.67	1.37	4.71	3.05	3.190	$p < 0.05$
S2 omission	1.08	0.79	2.57	2.90		
S2 addition	0.58	0.90	1.57	1.55		
S2 confusion	0.00	0.00	0.57	0.85		
S3 (parallel)	2.50	1.62	2.07	1.54	0.128	N.S.
S4 (free)	0.17	0.58	0.14	0.36	0.128	N.S.

N.S. = Not Significant.

The mean number of correct solutions and mean times to produce correct solutions at each level of complexity are summarized in table 5. The complexity of redescriptions was calculated in terms of the number of embedded elements in participants' actual solutions rather than in terms of the prescribed complexity of the sequence. This was because, in some cases, participants produced semantically correct redescriptions at different levels of complexity. Participants using the conditional notation produced more correct level 1 and level 2 solutions than participants using the disjunctive notation. Participants using the conditional notation also produced level 1 solutions faster (conditionals 142.2 s, disjunctives 212.1 s), $t = 2.505$, $p < 0.05$. There were no other significant differences between Notation groups. However, all participants got 100% of level 3 solutions correct regardless of notation used. This ceiling effect could be masking any differences between the two groups at this level of complexity.

Participants using the disjunctive notation typically made errors when they tried to insert other elements into the S2 element, where the separation between the two 'halves' of the S2 element was increased. This increases the processing load on the participant: s/he must remember what to put in the second half of the S2 element while filling in the first half. The conditional notation reduces this load: the *If...then...If not...then...* structure might 'cue' the participant as to what to put into each part of the element, whereas the *Either...then...or...then...* structure does not indicate clearly what should be put into each section. A similar account is offered by the 'parsing-gnisrap' theory of computer programming Green *et al.* (1987). Programming, they argue, involves two processes: parsing, the generation of external code structures from internal cognitive structures, and gnisrap (the reverse of parsing), which involves the recreation of internal structures from code during code comprehension. Different features of programming notations support each process. In the case of the SGT scheme it seems likely that users' cognitive structure of the sequences of actions are in terms of conditionals. It is therefore easier for them to create external structures for these sequences with a conditional than a disjunctive notation.

One counter-intuitive finding was that participants produced more correct solutions at higher levels of complexity, regardless of the notation used. There are two possible explanations for this finding. It could be that participants expend more effort on complex problems and, as a result, are less likely to make errors.

Table 5. Percentage of correct solutions and mean solution times (in seconds) at each level of complexity in experiment 1.

Level of complexity	Conditional notation		Disjunctive notation			
	Mean	SD	Mean	SD	t	Significance
1 embedded task	66.7	24.6	39.3	33.7	2.596	$p < 0.05$
	142	55	212	104	2.260	$p < 0.05$
2 embedded tasks	81.9	19.4	54.4	24.0	3.347	$p < 0.01$
	165	66	186	67	1.194	N.S.
3 embedded tasks	100.0	0.0	100.0	0.0		
	163	65	151	25	1.378	N.S.

N.S. = Not Significant.

Alternatively, the increased nesting in complex problems may help participants in some way. Sime *et al.* (1977c) and Green (1977) suggest that nesting helps in extracting sequence information from a description of the conditions under which actions can occur. If participants invested more effort in the more complex redescriptions then they would be expected to spend longer on these problems. However, the solution time data indicate that participants did not take longer to produce the more complex redescriptions. It seems instead that performance is better on the more complex redescriptions because the increased nesting helps participants to express the sequencing information implicit in the task descriptions.

5. Situating the SGT method within a computer-based tool

Although the results of experiment 1 show that the new version of the S2 sequencing element was much easier to use, an unacceptable number of mistakes were still made. For example, participants using the conditional S2 sequencing notation redescribed only just above one-half of the sequences correctly on the first attempt (admittedly, without having undergone training in using the SGT method beyond reading the experimental instructions). Thus, further development of the SGT method was clearly required. The chosen solution was to construct a computer-based environment within which the SGT scheme can be used.

There are several reasons why a computer-based tool may make the SGT method easier to undertake. First, Green (1989) argues that the environment in which a design methodology is used should provide the necessary features to support the use of the notations embodied within that methodology. Second, a computer-based tool can provide a formalism for applying the SGT scheme. Parnas and Clements (1986) describe several advantages of imposing order on the design process; for example, the designer can be guided through complex tasks to ensure accuracy and completeness (see also Macaulay *et al.* 1990), and progress can easily be measured and design projects consequently managed more efficiently. Third, tools allow the methodology user to handle large amounts of information easily, reducing design time and cost (Farooq and Dominick 1988). In particular, tools can be used to trap errors or check the completeness of the finished specification (Tse and Pong 1991). Finally, they can introduce consistency between users and end-products (Farooq and Dominick 1988).

Sime *et al.* (1977b) investigated how syntactic errors could be reduced when programming in NEST languages. Most syntactic errors involved the omission of 'END' statements marking the end of a conditional's scope. They found that participants using an editor that automatically provided conditional constructions (e.g. IF juicy THEN BEGIN xx END ELSE BEGIN xx END) solved more problems at the first attempt than participants who were taught a procedure for constructing programs. Parallels between programming and task redescription suggest that similar benefits might be obtained by providing automatic syntax generation in an SGT tool.

A computer tool was constructed to guide users through the process of using the SGT scheme using MacProlog32^{TM} (see Richardson (1996) for a full description). The tool has two main components: a Goal Information Window in which information on a task can be recorded and a Plan Construction Window in which a plan can be drawn up using the sequencing elements. The dialogue associated with the Goal Information Window, illustrated in figure 5, controls the design cycle underlying the SGT method. When it is first opened, the window contains a number and a label relating to the current high-level goal, an edit field containing a description of the current sequence and brief instructions. There is

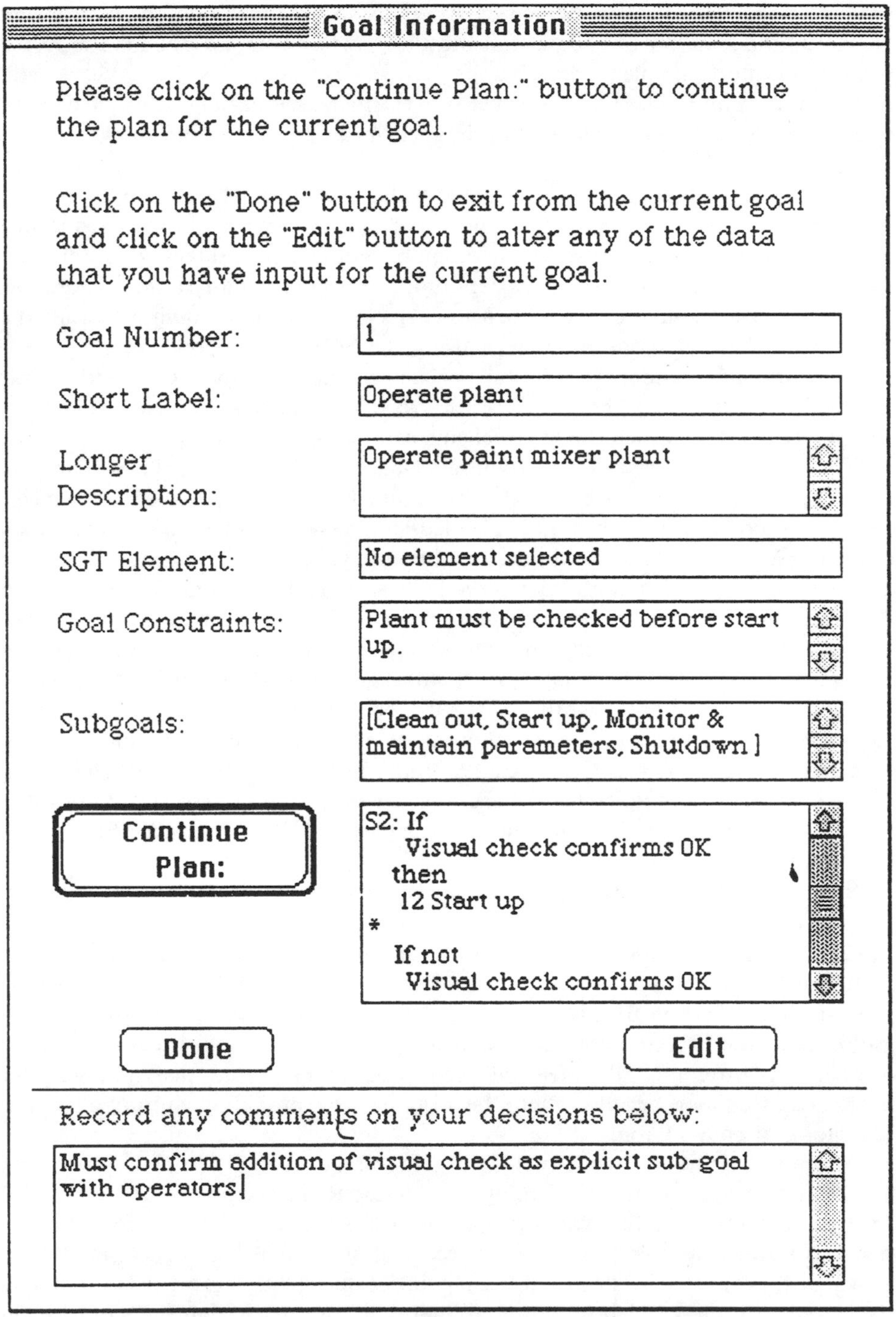

Goal Information

Please click on the "Continue Plan:" button to continue the plan for the current goal.

Click on the "Done" button to exit from the current goal and click on the "Edit" button to alter any of the data that you have input for the current goal.

Goal Number: | 1 |

Short Label: | Operate plant |

Longer Description: | Operate paint mixer plant |

SGT Element: | No element selected |

Goal Constraints: | Plant must be checked before start up. |

Subgoals: | [Clean out, Start up, Monitor & maintain parameters, Shutdown] |

Continue Plan:

S2: If
     Visual check confirms OK
   then
     12 Start up
 *
   If not
     Visual check confirms OK

Done   **Edit**

Record any comments on your decisions below:

Must confirm addition of visual check as explicit sub-goal with operators.

Figure 5. Goal Information Window in the SGT tool.

also a button labelled 'Subgoals...'. Clicking on this button, the user then records the sub-goals of the current high-level goal. Once s/he has done this, a button labelled 'Plan...' appears and the instructions change to tell the user the function of this new button.

Clicking on the 'Plan...' button opens a Plan Construction Window, illustrated in figure 6. The window contains a field in which the plan is written, a menu of sequencing elements, a field displaying information about the currently selected sequencing element and a menu of sub-goals. The information displayed for each of the sequence elements is shown in the Appendix. The user constructs the plan by selecting a point in the construction field then selecting an element or a sub-goal from one of the menus and clicking the 'OK' button next to the menu.

The Plan Construction Window prevents errors in two ways. First, whenever the user tries to insert an item into the plan, the validity of the insertion point is checked: for example, an error message appears if the user tries to put a sub-goal outside a

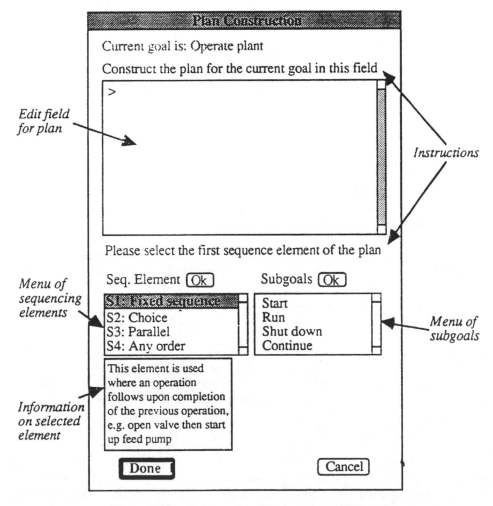

Figure 6. Plan Construction Window in the SGT tool.

sequencing element template. If the selected point is valid, then the chosen item is written at that point in the construction field. Second, a template for each selected sequencing element is written into the edit field with spaces for the insertion of subgoals. If the sequencing element is S2, S3 or S4, then a dialogue appears before the template is written to the plan construction field, to specify the conditions on which the action is dependent, or to specify how many sub-goals are linked by the element. When one sequencing element is placed within another, it is indented to make their relationship visually explicit. When the plan is completed the user clicks on the 'Done' button, and the plan to be written to plan field in the Goal Information Window. Input can subsequently be changed using the Edit function by altering, adding or deleting sub-goals, or by constructing a revised plan using a display of the original plan as a guide.

The SGT tool should reduce errors in two ways: first, the templates should prevent users from missing out parts of the elements or adding in unnecessary items and, second, the tool automatically checks the validity of the insertion point when a new item is added to the plan. Thus the tool has been designed specifically to support the use of the SGT notation and to prevent the types of errors that participants made in experiment 1.

6.   Experiment 2: evaluation of the computer environment

The aim of this experiment was to evaluate the usability of the tool against a paper-based environment for using the SGT scheme in task redescription. The experiment used the redescription problems from experiment 1, and the same performance measures were taken.

6.1. *Method*

6.1.1. *Participants*:   12 undergraduate students from the Psychology Department at Lancaster University were paid £5 each to participate in the study. None had taken part in experiment 1 and none had experience of programming, process control or engineering design. These participants were therefore from the same population as those who took part in experiment 1.

6.1.2. *Materials, design and procedure*:   The problems and experimental design were identical to those of experiment 1, except that the between-participants factor in this experiment was Environment, with two levels (Computer versus Paper). The procedure was also the same, with the exception of details on use of the computer tool that were added to the instructions.

6.2. *Results and discussion*

Participants' solutions were scored using the same criteria as in experiment 1. Table 6 summarizes differences in performance between computer and paper groups across a number of measures. No significant effect of Environment was found on the number of problems correctly solved by each participant. However, participants using the computer environment solved more problems correctly on the first attempt (4.5 correct) than participants using the paper environment (3.3 correct). Perhaps more importantly, significantly fewer errors were made by participants using the computer environment (1.8 errors) than by participants using the paper environment (4.6 errors). More specifically, participants using the computer environment made significantly fewer syntactic errors (0.3 errors) than participants using the paper

Table 6. Means and *t* test results for measures of accuracy, number of errors and solution time (in seconds) in experiment 2.

Measure	Computer		Paper		*t*	Significance
	Mean	SD	Mean	SD		
% Correct solutions	79.2	16.1	75.0	19.5	0.572	N.S.
% Correct on first attempt	75.0	16.7	55.0	22.8	2.382	$p < 0.05$
Total errors	1.83	1.27	4.58	3.28	2.703	$p < 0.05$
Syntactic errors	0.33	0.65	2.25	1.71	3.642	$p < 0.05$
Semantic errors	1.50	0.91	2.33	1.92	1.359	N.S.
Time taken for correct solution	252	34	157	65	5.638	$p < 0.05$
Error lifetimes	105	54	68	41	Not	analysed

N.S. = Not Significant.

environment (2.3 errors). No significant effect of Environment was found on the number of semantic errors.

As predicted, use of the computer tool reduced syntactic errors, while the number of semantic errors was not changed. Thus, the computer tool does not appear to affect how the participant understands the problem but it does help him/her to write his/her interpretation down correctly. There might be a limit to performance on this task, since participants in both conditions achieved the same number of correct solutions after error correction. Most of these failures were observed with the same problem, which required the nesting of a free sequence of sub-tasks within a contingent sequence which was itself nested within a parallel sequence. This is a complex sequence to interpret, and this difficulty was reflected in the semantic errors made by participants using both environments. Nonetheless, participants using the computer environment were more likely to produce a correct solution to this problem at the first attempt.

The mean times taken to produce correct solutions are also summarized in table 6. Participants took significantly longer to produce correct solutions using the computer environment than the paper environment. This is despite the fact that participants using pen and paper produced fewer correct solutions on the first attempt. This may reflect a speed-accuracy trade-off: participants using the computer environment might have been more accurate simply because they took longer. The dialogue for editing plans essentially required that a plan be started anew for an error to be corrected in the computer environment, while participants using the paper environment could correct mistakes without restarting the plan. It took longer to correct mistakes using the computer environment than the paper environment (although the error lifetime mean for the computer environment group is based on only 4 errors). Thus, it is possible that participants using the computer environment took more care over each plan.

In addition, participants using the computer environment might have taken longer because they not only had to draw up plans but also had to learn how to use the tool. If participants had been allowed to practice using the tool until they were completely familiar with it then the difference in times might not have been seen. However, both groups of participants were given only one practice trial so that they would have the same amount of experience with the SGT scheme. Figure 7 shows the change in solution times over the six trials. Participants using the paper environment

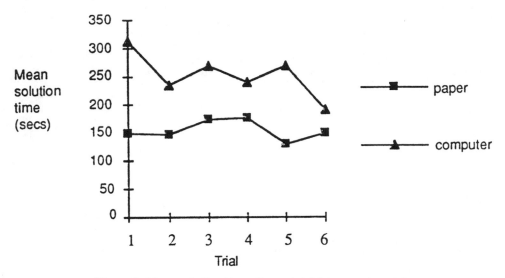

Figure 7. Mean solution times for each trial in experiment 2.

show only a slight decrease in solution times, while participants using the computer environment improved from 312 s on the first trial to 190 s on the sixth trial. This final figure is closer to the mean time of 150 s on the sixth trial for participants using the paper environment. This indicates that it is likely that some of the additional time taken by participants using the computer environment was spent on familiarization with the environment itself.

There are differences between the computer and paper environments in addition to the support of sequencing notations that might have led to qualitative differences in the task faced by each group. In particular, the computer environment required participants to restart plans from scratch when errors were made, it required participants to draw up the plan in a strictly sequential fashion and it involved many more steps than the paper environment. The longer solution times shown by the group using the computer environment may not simply be a tool-learning time, but may also reflect the fact that the computer tool made the redescription more complex and constrained than using pen and paper. However, the two groups did not differ in terms of semantic errors, which one might expect if the computer-based system changed the task fundamentally.

7. General discussion

The experiments demonstrate the importance of ergonomic design in developing a notation and environment for applying task-analysis methods. Experiment 1 showed that participants were able to use the new version of the S2 element (*if...then...if not...then...*) to express contingent sequences more accurately and quickly than they could with the old version (*either...then...or not...then...*). The results are in accordance with those of Evans (1982), who reports that negated disjunctives are rarely used in natural language. They also generalize the findings of Sime *et al.* (1977c) and Green (1977) concerning the NEST-INE notation to a domain other than computer programming.

Experiment 1 demonstrates the importance of designing a notation that is usable by participants and that supports the tasks for which it is used. In the case of the S2 element, the new version is easier for participants to understand and use and maps more closely onto their existing cognitive structures (Green *et al.* 1987). Obviously these two factors are inter-related: a notation that maps closely onto a cognitive structure will be easier to use. However, in some cases there might be several logically equivalent structures that all map onto the user's cognitive structures: in these cases the usability of individual forms will be more important.

In experiment 2, more error-free solutions and fewer syntactic errors were made by participants using the computer environment than a paper environment for redescribing plans using the SGT scheme. The provision of templates for SGT elements and automatic checks on the validity of insertion points in plans appears to facilitate the application of the SGT scheme. These results indicate how different environments constrain the ways in which a notation is used. For example, the paper environment allows participants to make changes and move quickly between different parts of a plan. It was therefore quicker to use initially than the computer tool, which restricted changes by the order in which parts of the dialogue were revealed. However, after six trials participants using the computer environment were equally as fast as participants using the paper environment. Thus experiment 2 demonstrates the importance of designing support tools for task analysis methods that specifically support the use of the notations embodied within the method. For example, by carefully studying the types of errors made by participants in experiment 1 it was possible to design a tool to prevent these errors.

The two experiments reported in this paper have also shown the limitations of syntax and environment in improving performance. In particular, both factors were shown to reduce only syntactic errors and not semantic errors. By improving the environment and the notation, the user can be helped to write down his/her interpretation more accurately by using the sequencing elements: however s/he cannot be helped to make a better interpretation. A similar finding was reported by Sime *et al.* (1977a) who found that templates reduced only syntactic errors in writing computer programs. It should be noted that experiment 2 does not, in itself, constitute a complete evaluation of the SGT tool, for two reasons. First, a number of important features of the tool were not evaluated, including the provision of HTA hierarchical displays from which task information can be accessed, and the automatic addition of stereotypical information requirements for each task to an HTA tabular display. Second, the success of the tool, and of the SGT method itself, can only be fully evaluated through its use on a realistic large-scale project involving HTA. Richardson (1996) reports the successful use of the tool by one expert designer working on a full HTA for the mixer plant shown in figure 1. However, further evaluation is being undertaken with deliberate caution, since it would be unfortunate for a potentially strong task-analytic method to be rejected as unusable, simply because the notation and support tools through which the method is delivered have not received sufficient attention.

It is important to design a notation and environment for a usability method that support the tasks for which the method is used. However, many methodologies use notations that have not been evaluated and tools that have not been carefully designed to support the use of the notation (see Richardson (1996) for examples where methodologies would have benefited from more careful notational design). There are other examples in the literature, in which careful evaluation of notations demonstrates considerable advantages. For example, Greene *et al.* (1990) compared the industry-standard database query language SQL, whose notation includes 'and' and 'or' statements as well as parentheses and keywords, with an alternative notation TEBI that does not have any of these notational features. Greene *et al.* (1990) found that participants using TEBI performed much better on a database querying task than participants using SQL. This is an important result, given the ubiquity of SQL. Similar importance should be attached to the ergonomic design of notations and tools for task analysis if it is to become more widespread as a design method, especially for non-specialist use.

Acknowledgements

The SGT method was originally developed by the first and third authors as part of a research contract for British Gas plc. The experiments and tool development were conducted as part of the second author's PhD research, which was supported by a studentship from British Gas plc. A short report of these experiments appears in S. Howard, J. Hammond and G. Lindgaard (eds), 1997, Human-Computer Interaction: Interact '97, *Proceedings of Interact '97*, Sydney, August 1997 (London: Chapman & Hall).

References

ANNETT, J., DUNCAN, K. D., STAMMERS, R. B. and GRAY, M. J. 1971, *Task Analysis* (London: HMSO).

BENYON, D. 1992, The role of task analysis in systems design, *Interacting with Computers*, **4**(1), 102–123.

CARD, S. K., MORAN, J. P. and NEWELL, A. 1983, *The Psychology of Human-Computer Interaction* (Hillsdale, NJ: Erlbaum).

EVANS, J. ST. B. T. 1982, *The Psychology of Deductive Reasoning* (London: Routledge and Kegan Paul).

FAROOQ, M. U. and DOMINICK, W. D. 1988, A survey of formal tools and models for developing user interfaces, *International Journal of Man-Machine Studies*, **29**, 479–496.

GREEN, T. R. G. 1977, Conditional program statements and their comprehensibility to professional programmers, *Journal of Occupational Psychology*, **50**, 93–109.

GREEN, T. R. G. 1989, Cognitive dimensions of notations, in A. Sutcliffe and L. Macaulay (eds), *People and Computers V* (Cambridge: Cambridge University Press), 443–460.

GREEN, T. R. G., BELLAMY, R. K. E. and PARKER, J. M. 1987, Parsing-Gnisrap: a model of device use, in F. M. Olson, C. Sheppard and E. Soloway (eds), *Empirical Studies of Programmers: Second Workshop*, (Norwood, NJ: Ablex).

GREENE, S. L., DEVLIN, S. J., CANNATA, P. E. and GOMEZ, L. M. 1990, No IFs, ANDs, or ORs: a study of database querying, *International Journal of Man-Machine Studies*, **32**, 303–326.

HARKER, S. D. P., OLPHERT, C. W. and EASON, K. D. 1990, The development of tools to assist in organisational requirements definition for information technology systems, *Human-Computer Interaction – INTERACT '90* (Amsterdam: Elsevier), 295–300.

LANSDALE, M. W. and ORMEROD, T. C. 1994, *Understanding Interfaces: A Handbook of Human-Computer Interaction* (London: Academic Press).

LIM, K. Y. and LONG, J. 1994, *The MUSE Method for Usability Engineering* (Cambridge: Cambridge University Press).

MACAULAY, L., FOWLER, C., KIRBY, M. and HUTT, A. 1990, USTM: a new approach to requirements specification, *Interacting with Computers*, **2**, 92–118.

PARNAS, D. L. and CLEMENTS, P. C. 1986, A rational design process: how and why to fake it, *IEEE Transactions on Software Engineering*, **12**, 251–257.

PAYNE, S. J. and GREEN, T. R. G. 1986, Task-action grammars: a model of the mental representation of task languages, *Human-Computer Interaction*, **2**, 93–133.

RICHARDSON, J. 1996, Cognitive processes in requirements specification, unpublished PhD thesis, Lancaster University.

RICHARDSON, J. and ORMEROD, T. C. 1997, Rephrasing between disjunctives and conditionals: mental models and the effects of thematic content, *Quarterly Journal of Experimental Psychology*, **50A**, 358–385.

RICHARDSON, J., ORMEROD, T. C. and SHEPHERD, A. 1998, The role of task analysis in capturing requirements for interface design, *Interacting with Computers*, **9**, 367–384.

SCHAFTEL, 1994, Personal Communication.

SHEPHERD, A. 1986, Issues in the training of process operators, *International Journal of Industrial Ergonomics*, **1**, 49–64.

SHEPHERD, A. 1990, Operating the mixer plant, Unpublished course notes, *Human Factors in Reliability*, TUDelft, Loughborough University, Department of Human Sciences.

SHEPHERD, A. 1993, An approach to information requirements specification for process control tasks, *Ergonomics*, **36**, 1425–1439.

SHEPHERD, A. and ORMEROD, T. C. 1992, Development of a formal method of user requirements specification for process plant displays. Report, Loughborough University.

SIME, M. E., ARBLASTER, A. T. and GREEN, T. R. G. 1977a, Reducing programming errors in nested conditionals by prescribing a writing procedure, *International Journal of Man-Machine Studies*, **9**, 119–126.

SIME, M. E., ARBLASTER, A. T. and GREEN, T. R. G. 1977b, Structuring the programmer's task, *Journal of Occupational Psychology*, **50**, 205–216.

SIME, M. E., GREEN, T. R. G. and GUEST, D. J. 1977c, Scope marking in computer conditionals – a psychological evaluation, *International Journal of Man-Machine Studies*, **9**, 107–118.

TSE, T. H. and PONG, L. 1991, An examination of requirements specification languages, *The Computer Journal*, **34**, 143–152.

WEBB, J. 1996, The design of computer telephony dialogues, Unpublished MSc Thesis, Lancaster University.

Appendix

The information displayed to users of the tool for each of the sequencing elements.

Sequence element	Information
S1: Fixed sequence	This element is used where an operation follows upon completion of the previous operation, e.g. open feed valve then start up feed pump.
S2: Choice	This element is used where different circumstances lead to different courses of action. The outcome of the previous operation determines which operation should follow it, e.g. inspect the seals, if they are leaking then switch to standby.
S3: Parallel	This element is used where an operation is followed by two or more operations that have to be carried out together, e.g. start the pump, then at the same time set feed line and start drain pump.
S4: Any order	This element is used where an operation is followed by two or more operations than can be carried out in any order, e.g. check water availability, then in any order check the availability of high and low pressure steam.

CHAPTER 8

Structured task analysis in complex domains

PATRIZIA MARTI

Multimedia Laboratory, University of Siena, Via dei Termini 6, 53100, Siena, Italy

Keywords: Task analysis; Structured design methods; Air Traffic Control.

This paper describes the application of MUSE, a human factors structured design method, to the design of systems supporting the work of air traffic controllers at European airport control towers. The paper illustrates the derivation of a generalized task model starting from the analyses performed at four airports and the construction of the composite task model of the future system. Other important concerns of the work are the harmonization of the different tools in the airport control tower and the assessment of the impact of these tools on the human operators. The potentialities and limitations of applying the method are discussed in particular with respect to the complexity of the domain and the non-standard features of the air traffic control as a MUSE application. Indeed, air traffic management is widely acknowledged as a complex domain to design for, and the documented application of human factors structured methods to design are all referred to the modelling of worksystems (as opposed to working positions or workstations) with a single operator. In the present application one has multiple operators and multiple design goals (design, harmonization, assessment). The work presented is a contribution to the extension and better definition of structured methods and their applicability to the analysis of complex domains.

1. Introduction

Task analysis and task modelling are major human factors tools for interactive system design. The models they produce are means for understanding, communicating, testing and predicting fundamental aspects of the human-machine system being developed or evaluated. They help designers to acquire a user-oriented perspective of the activity, thus avoiding function-oriented biases during the development process (Kirwan and Ainsworth 1992).

The structured design methods based on task analysis integrate human factors and software engineering in the design process. This paper describes the application of MUSE (Lim and Long 1994), a human factors structured design method, to the design of air traffic control systems in European airport control towers. The air traffic management is widely acknowledged as a complex domain and the application of human factors structured methods to design is both relatively novel and seldom reported in the literature. The Method For Usability Engineering (MUSE) is supposed to overcome important criticisms concerning the applicability of task analysis (Lim 1996). These include: the requirement for an existing system, the focus on analysis rather than design, the limited scope of application in the design cycle, the under specification of the application domain, the inadequate documentation of outputs, the little guidance in the selection of a particular method.

The interest of the application concerns the complexity of the domain but also its non-standard feature as a MUSE application. Indeed the documented case studies of MUSE regard the modelling of worksystems (as opposed to working positions or workstations) with a single operator, such as recreational facilities booking (Middlemass *et al.* 1995) or domestic energy management (Stork *et al.* 1995). In the present application one has multiple operators and multiple design goals, namely the design of future air traffic control systems, the harmonization of the new advanced technological tools in the control tower and the assessment of the impact of these tools on the human operators.

In the following, an overview of MUSE and a description of the application domain is given. The paper illustrates the task analysis process, in particular the derivation of a Generalized Task Model (GTM) starting from the analyses performed at four airports, the construction of the composite task model of the future system, and their implications for the design. In the final discussion three critical aspects experienced in the application of the method are considered: generification and transition from the Generalised Task Model to the Composite Task Model; modelling of co-operation and communication; and modelling of the right unit of analysis.

2.   An overview of MUSE

The following is a description of the MUSE method (Lim and Long 1994). The method develops in three main phases: information elicitation and analysis phase, design synthesis phase and design specification phase.

The first phase of the method involves collecting and analysing information intended to inform later design activities. This phase is carried mainly using task analysis techniques. Information elicitation and analysis phase consists of two stages, the extant system analysis and the Generalised Task Model stage. During the former, background design information is collected that relates both to the system currently in use and other related systems, for example systems having a similar task domain. The information concerns the user needs and problems, existing task characteristics, design features and rationale, the devices used and the task performed. Its main objective is to identify problems that can be overcome in the design of the new system. The GTM stage provides a device-independent task model of the existing system.

The second phase of MUSE, the design synthesis phase, begins by establishing the human factors requirements of the design, and collects them in statements of user needs. Then, a Composite Task Model (CTM) is developed, which expresses the conceptual design of the target system. It is divided in system and user task model to separate the sub-tasks that are to be performed using the system under development from those that are performed using other devices. In this model, tasks are allocated to the user or the system or both of them (mixed tasks). The system task model describes also the target system functionality.

The third phase of MUSE, the design specification phase, provides a device-specific implementable specification. It consists of two stages, the interaction task model stage and the display design stage. The interaction task model provides the specification of device-level actions to be performed by the user, the specification of display objects, their behaviour and appearance; the display design stage provides a set of pictorial screen layouts corresponding to the specification provided by the interaction task model. This defines the context for activating error and feedback messages, specification of the content and layout of each display screen.

3. An application of MUSE to air traffic control

In the following the application of MUSE to Air Traffic Control (ATC) is described. The work has been carried out within the European project ATHOS (Airport Tower Harmonised cOntroller System—TR1005 of the ECARDA Programme, CEC DG XIII) started in 1996 and still underway. The main goal of ATHOS is the design of future airport tower control systems. It deals with the tower controller positions in visual control rooms in the context of future SMGCS (Surface Movement Ground Control System). ATHOS moves steps towards the European ATC harmonization and the design of the controller working positions (CWPs) of the 'standard European Tower Control'.

Specific objectives of the project are as follows.

(1) Design of a mock-up for the future airport tower control system, in particular for the ground and tower controller working positions.
(2) Harmonization in the control room of the new technologies for ATC that will be operational in the next 10–15 years. These include:
    (a) air/ground data-link and gate-link between controller and pilot;
    (b) planning tools: automatic conflict detection, departure manager, arrival manager;
    (c) data fusion on mixed data sources (radar, beacons, etc.);
    (d) short-term accurate meteorological and wake vortex forecasts; and
    (e) advanced HCI technology concerning the display functions and the human-computer interaction.
(3) Assessment of their impact on the human operators.

The rationale for the selection of MUSE in ATHOS can be summarized in the following points.

(1) The need of a method supporting human factors throughout the whole development process.
(2) The need of a method that integrates human factors and software engineering methods.
(3) The need of a structured development process involving a broad view of task analysis, ensuring the effective exploitation of task analysis results throughout the process.
(4) The existing situations analysed: the different airports show significant similarities, but the detailed structure is very dependent on the work organization, the topology of the airport, the traffic load, the tools used, local regulations, etc. In this respect, the abstraction from the current systems with the aim of a device-independent analysis is an interesting aspect of the methodology.
(5) The difficulty in assessing the impact of new technological tools on the activity, starting from different practices and habits.
(6) The objective of the standardization of controller working positions, and the assessment of controllers' acceptance of the new organization.
(7) The heterogeneity of tasks, the complexity of the data involved, the high degree of parallelism and cycles during the activity.

Four important European airports were involved in the analysis activities: Amsterdam Schiphol; Paris Charles De Gaulle; Frankfurt/Main; and Fiumicino.

For each airport, all the CWPs were modelled, even if this focused on Ground and Tower for the development of the final mock-up.

The task analysis technique that was used was inspired by MAD (Scapin and Pierret-Golbreich 1989) and TKS (Johnson and Johnson 1991). From TKS the philosophy of the approach and the graphical notation was taken. From MAD the use of templates conveying information about tasks not contained in the task graphs was taken. The basic element of the task analysis is a graphical notation to draw task graphs, i.e. a task taxonomy representing the goal structure of tasks. This structure conveys a task/sub-task decomposition, defined in terms of temporal and logical constraints among tasks; triggering conditions; loops, etc. Pre-conditions and post-conditions for task execution are specified when necessary, as well as pre-requisite and post-requisite, which are tasks that need to be executed before or after a main task.

Each task in the graphs is also described in a task template collecting functional, operational, pragmatic and relational characteristics of the task. Other information about non-functional requirements of tasks such as duration, interruptibility, reversibility, frequency are also specified as natural language descriptions in the task template. These aspects of the task analysis approach are inspired by MAD.

Other relevant elements of the task modelling are:

(1) descriptions of roles, i.e. a set of tasks that each user is responsible for performing as part of his/her duties in the context of a job; and
(2) descriptions of objects, i.e. a set of objects or classes of objects involved in the tasks, and the semantic relationships among them.

Both roles and objects are described in role and object templates containing information about relations among roles, objects' data attributes, and actions that may be applied to the objects. This object-oriented aspect of task analysis adopted was the natural link with the object-oriented analysis method used later on in the project, therefore with the software engineering phase.

4. The complexity of the ATC domain

The control tower of an international airport is characterized by a work organization that is strictly co-operative in nature. The general goal of the control tower—shared by all controllers—is to ensure a safe and expeditious traffic flow within the aerodrome and from/to subsequent areas of responsibility (Hopkin 1995).

Tasks and duties are precisely distributed among several controllers (5–10 controllers) within defined areas of responsibility, and controlled objects are handed over from a controller to the subsequent one according to defined procedures. Communications and co-ordination play a key role in the successful management of tasks. Indeed, each controller needs to be informed on the actual and planned traffic situation in adjacent areas and on the control strategies being applied by other controllers (situation awareness). In this way the controller is able to anticipate over some minutes the evolution of the traffic situation within his own area, to identify in advance conflicts, and to help other controllers to face critical situations. All the activity is therefore characterized by both explicit communications (direct talk, transmission of flight progress data) and a background activity of monitoring adjacent controllers.

This aspect is relevant particularly to the work of GC (Ground Controller) and TC (Tower Controller), who in many cases need to act in strict co-ordination for handling traffic flow in proximity of runways. This means that although responsibilities over moving aircraft are clearly defined—it is not possible that an aircraft is simultaneously under control of both GC and TC—the physical areas monitored by the two controllers are overlapping.

Several factors contribute to increase the complexity of the controller's activity to a critical level:

(1) heterogeneity of tasks (assessment, guidance, monitoring, control, planning, communications);
(2) heterogeneity of tools: the execution of the same task can be supported by different tools, possibly chosen according to the current context (e.g. the use of monitoring tools is dependent on visibility conditions);
(3) multi-tasking: concurrent handling of different tasks or different instances of a task (e.g. many aircraft controlled at the same time) is the general rule of controllers. This feature is highlighted in the task model by the frequent use of interleaving and iteration;
(4) variability in external conditions: environmental data (e.g. weather data) must be assessed continuously to ensure best and safest use of available resources;
(5) time pressure;
(6) constraints to be satisfied (aircraft constraints, several types of regulations, priorities, meteorological conditions);
(7) scarce or no support used for keeping a picture of the situation, in terms of planned traffic (taxi routes, acceptance rate of controlled and subsequent areas, sequencing), specific situations that may cause conflicts, configuration of resources (e.g. runway use). Most of such information is actually stored in the memory of the controller;
(8) concurrent monitoring of both specific situations (sets of controlled objects, either co-operating or not co-operating) and overall area of responsibility;
(9) continual effort spent in anticipating the evolution of the current situation;
(10) noise produced by telephone, radio and hand signals; and
(11) the impact of potential errors.

5.   The task analysis process

In what follows, the task analysis process adopted in ATHOS is described. Since it is desirable to discuss limits and potentialities of the approach in particular during analysis and conceptual design, the focus will be on the first two phases of the MUSE approach: the information elicitation and analysis phase and the design synthesis phase. the task analysis process is illustrated as it was applied to one of the case studies considered in the project: the design of the tower controller working position.

Using the notation illustrated above, several task models were produced during the extant systems analysis, constituting the basis for the design synthesis phase. The task models produced served to represent the specificity of the different situations analysed and were elaborated to produce a generalized task model. This provided synthesis and abstraction of the different task models developed in the four airports, highlighting the common general structure and representing the specificity as possible elements of the activity in general.

The GTM was a fundamental step towards the design activities; indeed it allowed the production of a model that is an abstraction from the current systems with the aim of a device independent analysis of the activity.

Task modelling was an iterative process: at each task analysis stage, the models were validated with controllers, and a final evaluation on the GTM was performed on the basis of the compliance of the model with the current activities in the different airports, taking into account completeness, accuracy, expressiveness of the model, representativity of the activities, resulting scope of the task model.

Furthermore, task modelling was a participatory design process because the controllers involved in the project actively contributed in the extant systems analysis, and the construction of the GTM. At the beginning of the project, air traffic controllers and designers attended a course on the task analysis process and the notation adopted, to enable them to refine and validate the models.

5.1. *The generalized task model*

This section gives a flavour of the GTM for the TC's activities. Indeed the complete GTM for all the CWPs is a large document, quite complex and difficult to manage. Figure 1 gives the basic elements of the notations used.

The main goal of the activity of the TC is to make the best possible use of runways for aircraft landing and take-off while maintaining high levels of security. The most important requirements of the TC are related to automatic tools supporting planning activities, in particular runway management and aircraft sequencing.

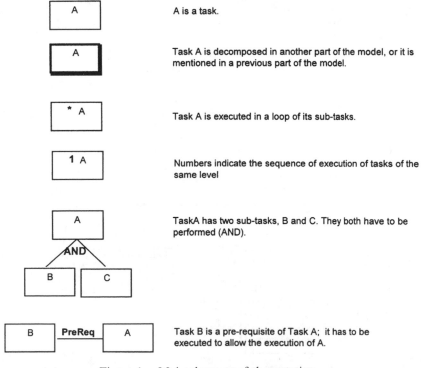

Figure 1.   Main elements of the notation.

The TC is responsible for the control of aircraft in the traffic circuit, i.e. the landing aircraft until they have vacated the runway and the departing aircraft from the holding position for the take-off until they are handed over to APP (Approach Controller).

The activity of the TC consists of five main tasks: monitoring, planning, control, guidance, co-ordination/communication, and additionally, the configuration of runways (figure 2). The main features of these tasks are described below and the 'plan traffic management' task is discussed (figures 3 and 4).

5.1.1. *Monitoring*:   The monitoring task of the TC is for the greater part based on surveillance by visual inspection. The ordering of flight strips indicates how the current situation is organized. The monitoring task is necessary to form a mental picture of the actual traffic situation and to check whether the aircraft are moving according to their plans and their instructions. An important aspect of the monitoring function of the TC is the verification that the runway is vacated.

5.1.2. *Control*:   Control concerns the detection and solving of traffic conflicts and plan deviations. The runway area is highly critical. The separation rules are

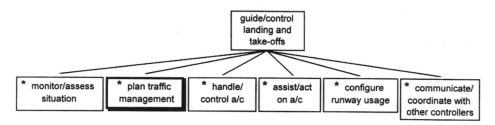

Figure 2.   The top level task model for the TC.

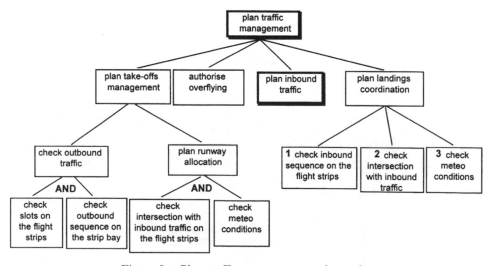

Figure 3.   Plan traffic management task graph.

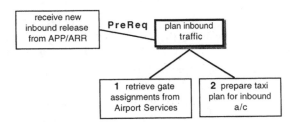

Figure 4. Plan inbound traffic task graph.

compared with the actual traffic situation and orders are given to respect the rules. Wake vortex is an important factor in applying separation. For the control task it is necessary to anticipate on the traffic situation of a few minutes ahead.

5.1.3. *Guidance*:   The TC clears inbound aircraft for landing and outbound aircraft for take-off. When necessary, additional information is supplied. The TC gives additional route information to outbound flights. Communications to pilots are handled via radio frequency. Guidance may also be given via airfield lighting.

5.1.4. *Co-ordination/communication*:   Co-ordination and communication is mainly necessary with the GC, the approach controller and the supervisor.

5.1.5. *Configure runway*:   The runway configuration task concerns the establishment of the best configuration for runway use, according to meteorological conditions, runways operational status and taxiway configuration.

5.1.6. *Planning*:   All traffic movements including arrivals, departures, and VFR over-flight traffic are planned by the TC. Based on the mental picture, the TC makes an initial traffic plan, which matches the constraints of the flight plans. The flight strips are sorted so that they are in a temporal ordered sequence. Clearances from the controller, reports from the pilot, and co-ordination with other controllers may affect the sorting of the strips. Usually, new strips are inserted at the end of the traffic queue, but when gaps in the sequence appear, the controller may in some cases decide, in co-ordination with other controllers, to fill these in with new aircraft. Replanning may be necessary when deviations occur.

5.2. *The Composite Task Model*
Starting from the GTM and the statements of user needs the CTM, a model of task allocation between the TC and the future system, was provided by the author. In this model, each sub-task is allocated either to the controller, or to the system, or is marked as a joint task between the user and the system. The model below strictly concentrates on the planning task of the TC, as evolution of the GTM presented above.

Additional elements were added to the initial notation to distinguish between user and system tasks. These elements are represented in figure 5.

The CTM shows at the high level many similarities with the GTM. The most significant differences between the two models are related to the automatic support to the 'check inbound traffic' and 'plan runway allocation' tasks. The display functions of slot, scheduled departure time, and outbound sequence are output of the automatic planning tools (currently under development), the *departure manager* and the *arrival manager*. The *arrival manager* is mainly targeted at the Approach Controller but delivers its results to the TC. He is in charge of planning the aircraft approach sequence. The departure manager integrates the activities of GC and TC during taxiing and runway occupancy providing the departure sequence at the runway entry points, with respect to the aircraft constraints (slot, flight plan, type of aircraft, etc.). The usefulness of these tools is widely acknowledged by the controllers, nevertheless it is necessary to design an effective external representation of the data in order to successfully exploit the potentialities of the tools. The problem is therefore to integrate these tools in the context of the activity of the TC, providing all information needed to execute the runway management task. For a presentation of the user interface developed for the general task of plan traffic management (plan take-offs management, plan runway allocation, plan inbound traffic, figures 6–8) see Marti (1997).

6. Discussion

In this section the application of MUSE to the ATC application presented in this paper is discussed. In particular, four critical aspects are considered: generification; transition from the GTM to the CTM; modelling of co-operation and communication; modelling of the right unit of analysis.

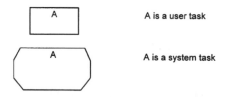

Figure 5.   Additional elements of the notation for CTM.

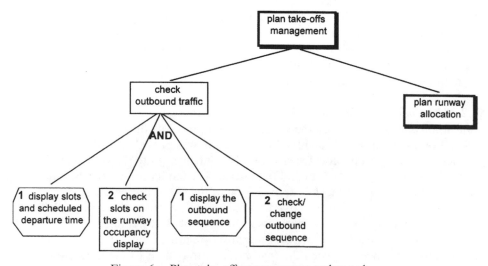

Figure 6.   Plan take-offs management task graph.

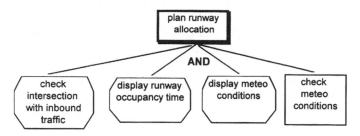

Figure 7. Plan runway allocation task graph.

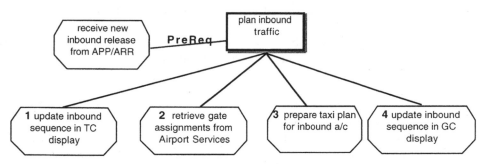

Figure 8. Plan inbound traffic task graph.

6.1. *Generification and transition from the Generalized Task Model to the Composite Task Model*

The GTM is a fundamental step towards the design activities. However, experience showed that the construction of GTM and the transition from the GTM to the CTM are not adequately supported in MUSE. For this reason, an attempt was made to define guidelines for generification and user-centred design techniques were integrated (Norman and Draper 1986) for supporting the transition from GTM to CTM. Both solutions have been successfully applied in the design of a decision support system for handling environmental emergencies (Marti and Normand 1997).

6.1.1. *Guidelines for generification*: This approach to generification develops in four main phases:

(1) *Integration and reorganization of the task graphs of the extant systems.* The integration and reorganization of the task graphs is a rather creative process seeking to minimize the modifications to the existing task structures reflected in the extant system models, and saving the peculiarities when relevant.

(2) *Updating of task description.* Updating of task description is necessary when task descriptions are missing or incomplete. In these cases, information coming from other sources should be used to refine the description. In particular, textual descriptions taken during the interviews and observations of the controllers at work should be used to complete the task templates and to clarify the information content. Other updating consists of a revision of

the task graph, for example when a new general task is defined by abstraction of some other existing tasks. Observation of the users at work is always necessary to updating the task description and filling in missing data.

(3) *Check of task graph consistency.* Check of task graph consistency is an important and time consuming activity. It includes the following.

    (a) Checking that a task description is complete and consistent. This point is very critical; indeed even if the task models are representative for the activities currently performed, complete consistency cannot be guaranteed. Another important aspect to keep in mind is the difficulty of maintaining consistency and completeness during the generalization. In many cases, producing consistent models means producing models that are incomplete with respect to the initial data. This limitation is unavoidable.

    (b) Checking that the defined roles correspond to the tasks modelled in the task graph.

    (c) Checking that the objects defined as domain objects are effectively being referred to at least one task description. The definition of the domain objects is very important because it forms the specifications of the interface design.

(4) *Check and validation with task performers.* Check and validation with task performers should be performed iteratively on the intermediate draft task models. Task scenarios (Carroll 1995) can be used to verify the strategies for task execution. They are represented by descriptions of realistic situations during which the users have to execute a task following their own strategies.

In the application presented in this paper, the guidelines for task model generalization were applied as described below. Starting from the top hierarchy the generalization was carried out by looking for similar tasks (mainly tasks that have the same goal), and checking to what extent they shared common elements, e.g. objects, procedures, or sub-task constituents.

When a group of similar tasks was identified and the comparison established, a superordinate task was defined. In some cases new generic task names were substituted for the former ones to allow the abstraction. In other cases when the same task was decomposed in different hierarchies of atomic sub-tasks, then the different structures were represented as alternative ways to execute the task. This was done to save the peculiarities of the models that can provide important insights during the design.

When it was established that the variations among similar tasks were not due to different task performance strategies but to some local external constraint, then they were removed in the task graph and (if relevant) reported in the task template.

Last, but not least, an important criterion for generalization was determined by the prevailing design context. In the present case, there were three main design objectives for the project: the design of a new CWP, the harmonization of the user interface of the different tools of the control tower, and the assessment of the impact of the new technological tools on the human operator. The prevailing design context was therefore the development of a new human-computer system on the basis of common requirements, common objectives, and different control tower regulation and organization. From this point of view the aim was to provide a GTM minimizing the differences in task performance related to the specific device used, and highlighting those aspects of the activity that require a computer-based support in the new system.

6.2. *Transition from GTM to CTM*

The conceptual design of the new system is expressed in the form of a CTM. The main problems of the transition from GTM to CTM are related to the description of the *future activity* (user/s interacting with the new system), and the *task allocation* (Sutcliffe 1995). Both of these activities cannot be adequately executed without an early use of prototypes and user involvement. Indeed the CTM is an interpretation of the designer of how the human-machine system should be to satisfy the initial requirements. Its construction can benefit from the application of user-centred design techniques and rapid prototyping to verify fundamental steps such as the task allocation between system and user. In MUSE, the prototyping of the user interface is left to a late stage in the design process, and this causes problems in fixing and redesigning defects of the conceptual design phase.

Furthermore, the process of building a CTM should be carried out as an iterative process: at each design stage, the model should be validated and refined with the task performers, taking into account completeness, accuracy, expressiveness and resulting scope of the task model. With the integration of user-centred design techniques it is possible to fully exploit the potential of the CTM which can support the prediction of potential usability problems of the final design and guiding redesign.

6.3. *Modelling co-operation and communication*

Even if many task analysis methods are currently available, they are still poorly developed to model communication tasks and co-operative activities among different actors. Also the documented case studies of MUSE all refer to single user applications. In ATC co-ordination and communication play a key role and in ATHOS the author experienced many problems in modelling both the aspects. The attempt that the author made to face the problem was to provide a description of the roles that the different agents in the control tower play during their activity. This description was provided as a set of tasks that each user is responsible for performing as part of his/her duties, and the relations among the roles. This set of tasks was matched on the task model, so that it was possible to understand which tasks were shared by the different controllers. The description of the roles was complemented by the description of the objects (including messages) involved in the task execution. In this way one had a three-dimensional view of the activity: the dimension of the tasks, the dimension of the roles sharing certain tasks and the dimension of the objects that are handled during a co-operative activity. The approach seemed interesting in its ability to inform the design. Indeed it provided information about the degree of support needed during co-operative activities, which was not simple to capture from direct observation of the activity or from interviews.

6.4. *Modelling the right unit of analysis*

Another problem of the approaches based on task analysis is the definition of the right unit of analysis. Cognitive science normally takes the individual agent as its unit of analysis as if all the interest in the human endeavour is determined by the information processing properties of the individuals. Recent approaches to distributed cognition offer a framework that takes a distributed, socio-technical system rather than an individual mind as its primary unit of analysis.

Hutchins (1995) describes the analysis of a memory task in the cockpit of a commercial airline. His work shows how the cognitive properties of such distributed systems can differ radically from the cognitive properties of the individuals who inhabit them. Hutchins takes the cockpit as the unit of analysis and demonstrates how the cockpit system performs the cognitive tasks of computing and remembering a set of correspondences between airspeed and wing configuration. Within the distributed cognition framework what is important during the analysis of the human activity is the interaction of people with each other and with the physical structure in the environment. The unit of analysis is larger than a person therefore it must be enlarged to include a system where the cognition is distributed among people and the external environment.

The relationship between distributed cognition and the existing task analysis methods is still unexplored. The problem of the unit of analysis is an important criticism that can be moved in particular to those task analysis methods modelling sequences of low level actions. These methods, such as GOMS (Card *et al.* 1983), TAG (Payne and Green 1986) and UAN (Hartson and Gray 1992) relate to the minute detail of how a user interacts with a computer at the command and input/output level, instead of detailing a higher unit of analysis.

7. Conclusions

This paper describes an application of the structured method MUSE to the ATC domain and discusses potentialities and open points. ATC is surely a non-standard application for MUSE. The documented case studies of MUSE refer to the design of working systems for single user applications. The present attempt was to try to use the method to design CWPs for teamwork. Other primary objectives of this work were the harmonization of the technological tools in the tower control (as part of the standardization process of the European airport control towers) and the assessment of the impact of these tools on the human operators. These objectives provided the rationale for the selection of MUSE as the design methodology: a method was needed for analysing the different existing situations in four important European airports and for providing a unified view of the activity. This unified view aimed at guiding the design of the CWPs. Starting from 'well defined observable user requirements' (in the terminology of MUSE) the structured approach allowed one to acquire a user-oriented characterization of the activity and to maintain a well-documented process where design decisions were traced by means of task model evolution. The task models produced were furthermore an effective communication means among designers who worked for each other.

The paper offers a concrete example of modelling activities in complex domains. This experience within a real-size project highlights the benefits of the structured design method but also reveals its limitations. In particular the practice of task analysis and its efficacy can be enhanced if some of the basic concepts and processes of MUSE are extended and defined more explicitly. These are essentially the generification and the transitions among the phases. A contribution to a better definition of the generification is given in this paper by the presentation of guidelines, a set of checklists and suggestions to carry out successfully this critical phase of the process. Since task analysis is a time-consuming activity, a guidance to the application of the method, in particular during the generification, can limit the employment of resources necessary to carry out the complete task analysis process.

With respect to the transition from GTM to CTM the paper recommends the integration of user-centred design techniques for involving the users in developing early prototypes: these can definitely contribute to improve the overall design process, and provide adequate guidance to take design decisions (in particular during task allocation between user and system). In MUSE the development of user interface prototypes is considered only in the third phase of the process, the design synthesis phase.

The paper offers points of discussion on two important problems: the modelling of co-operation and communication and the choice of the right unit of analysis. Both these aspects constitute a challenge for structured design methods based on task analysis.

References

CARD, S. K., MORAN, T. P. and NEWELL, A. 1983, *The Psychology of Human-Computer Interaction* (Hillsdale, NJ: Lawrence Erlbaum).

CARROLL, J. M. 1995, *Scenario-based Design* (New York: Wiley).

HARTSON, H. R. and Gray, P. D. 1992, Temporal aspects of tasks in the user action notation, *Human Computer Interaction*, **7**(1), 1–45.

HOPKIN, V. D. 1995, *Human Factors in Air Traffic Control* (London: Taylor & Francis).

HUTCHINS, E. 1995, 'How a cockpit remembers its speeds', *Cognitive Science*, **19**, 265–288.

JOHNSON, P. and JOHNSON, H. 1991, 'Task knowledge structures: psychological basis and integration into system design', *Acta Psychologica*, **78**, 3–26.

KIRWAN, B. and AINSWORTH, L. K. 1992, *A Guide to Task Analysis* (London: Taylor & Francis).

LIM, K. Y. 1996, Structured task analysis: an instantiation of the MUSE method for usability engineering, *Interacting with Computers*, **8**, 31–50.

LIM, K. Y. and LONG, J. B. 1994, *The MUSE Method for Usability Engineering* (Cambridge: Cambridge University Press).

MARTI, P. 1997, 'Assessing the impact of new technologies for Air Traffic Control: the human factors perspective', *Proceedings of Cognitive Science Approaches to Process Control*, Baveno (Rowa: IP-CNR).

MARTI, P. and NORMAND, V. 1997, Bridging software design and usability analysis through task modelling, in K. Varghese and S. Pfleger (eds), *Human Comfort and Security of Information Systems* (Berlin: Springer Verlag).

MIDDLEMASS, J., STORCK, A. and LONG, J. 1995, 'Applying a structured method for usability engineering to recreational facilities booking user requirements: a successful case study', *Proceedings of DSV-IS '95* (Wien: Springer Verlag).

NORMAN, D. and DRAPER, S. W. (eds) 1986, *User Centered System Design: New Perspectives on Human-Computer Interaction* (Hillsdale, NJ: Lawrence Erlbaum).

PAYNE, P. and GREEN, T. R. G. 1986, Task-action grammars: a model of the mental representation of task languages, *Human Computer Interaction*, **2**, 93–133.

SCAPIN, D. and PIERRET-GOLBREICH, D. 1989, Towards a method for task description: MAD, *Proceedings of the Work with Display Units Conference*, Montréal, Canada (North Holland: Elsevier).

STORK, A., LONG, J. and MIDDLEMASS, J. 1995, Applying a structured method for usability engineering to domestic energy management user requirements: a successful case study, *Proceeding of HCI'95*, 368–385 (Amsterdam: Elsevier).

SUTCLIFFE, A. G. 1995, *Human Computer Interface Design* (Basingstoke: Macmillan).

CHAPTER 9

Task and training requirements analysis methodology (TTRAM): an analytic methodology for identifying potential training uses of simulator networks in teamwork-intensive task environments

Robert W. Swezey†, Jerry M. Owens†, Maureen L. Bergondy‡ and
Eduardo Salas‡

†InterScience America, Inc., 312 East Market Street, Suite G, Leesburg,
VA 20176-4173, USA

‡Naval Air Warfare Center, Training Systems Division, Code 4961, 12350
Research Parkway, Orlando, FL 32826-3224, USA

Keywords: Simulator networking; Team training; Training analysis.

A methodology designed to identify potential application areas for use of networked simulations is presented. The technique, known as task and training requirements analysis methodology (TTRAM), has been independently applied to the analysis of numerous US military aircraft simulator networking requirements, and appears to effectively discriminate tasks that are prone to skill decay, that are critical to mission success, that require high levels of internal and external teamwork, and that require additional training support.

1. Introduction

Many military and commercial tasks are characterized by dynamic exchanges of information and resources among group and team members, and by co-ordination of task activities. Performance on such tasks can be impaired, not because individuals within the group lack requisite abilities, but because the team as a whole fails to properly co-ordinate their activities. Although team performance can sometimes be improved by pre-training individual team members and/or teams to employ teamwork skills (McRae 1966), programmes that purport to train teamwork skills have traditionally emphasized individual skill learning within team settings, and have failed to consider special teamwork requirements beyond these principles. As a result, team members are often well trained in specific individual competencies required in a teamwork setting (i.e. taskwork skills), but are unable to adequately perform tasks that require a teamwork emphasis, or must be transferred to a more complex situation (Dyer 1984, Swezey and Salas 1992).

The structure and composition of a team can vary greatly, ranging from a two-person crew to a force consisting of thousands of individuals and sub-groups, each of which becomes an integral part of a larger collective team. As individuals are added to a team, not only does the overall pool of resources increase, but so do the co-ordination demands associated with completing team tasks. In large-scale systems, decision making is often distributed over several individuals or teams, who are geographically separated, but who must co-ordinate to share information and resources. In some cases, team performance gains resulting from an increase in task-related resources (due to increases in team size) can be offset by co-ordination demands (Shaw 1976). This phenomenon, known as process loss (Forsyth 1983), is well documented in the literature and has been attributed to a lack of generic teamwork skills (Swezey *et al.* 1994).

While the effective functioning of individual teams is important, such teams rarely function in isolation; basic sub-groups often form part of a larger 'team' (or collective group), whose size often depends upon the type of mission or task. To function effectively as an integrated team, organizations often must train personnel to operate as part of larger extended 'teams.' Training systems that combine sub-group elements through simulation, for example, provide an opportunity to work as part of a co-ordinated whole towards a common goal. Such systems may be considered to be an integral component of adequate team training programmes (Salas and Cannon-Bowers 1997, Swezey and Llaneras 1997).

One salient factor that tends to interact with team performance appears to be the nature of the task performed by the team. Environments that have few interdependencies among tasks (i.e. low task complexity) are not likely to require high levels of communication and co-ordination among team members, and thus can be successfully performed by individual team members with little or no interaction. However, when team tasks are highly interrelated (character-ized by high task complexity) team members must co-ordinate and communicate effectively in order to complete the task successfully (see Dyer 1984, Streufert and Swezey 1986, Swezey and Salas 1992, for discussions of these issues). Obrien and Owens (1969), for example, found that group performance is predicted by general group member ability in co-ordination tasks (those in which group members contribute independently to group performance), but not in collaborative tasks (where all group members must work together interactively). Thus, although both taskwork skills and teamwork skills influence the level and nature of team performance, the relative contribution of each to overall performance may vary according to the demands of a particular task. It is important, therefore, to identify and understand the underlying teamwork demands posed by operational tasks.

A variety of methods for analysing team tasks exist (Levine *et al.* 1988); most emulate traditional task analytic frameworks geared towards describing individual tasks. Hierarchical analytic techniques, for example, have been used to decompose team missions into subsidiary support activities through the identification of mission functions (collection of tasks that serve a common purpose and support mission accomplishment), tasks (activities performed together to achieve a specific goal), and sub-tasks (activities that contribute to the accomplishment of a task). However, traditional task analytic techniques typically do not provide mechanisms for capturing requisite cues, conditions, and standards for team tasks and skills. Andrews *et al.* (1992), for example, reported that traditional task analytic techniques were inadequate in specifying team skill and task requirements (e.g. communication, co-ordination, and team decision-making) for an antisubmarine warfare team training device. As with their individual-oriented counterparts, team tasks may be analysed at multiple levels of description (e.g. behavioural description, behavioural requirements, task characteristics, etc.) using numerous approaches, utilizing scales to assess team member attribute requirements, scales applied to task demands, and scales applied to team member attributes. One example of such an analytic methodology is presented in this paper.

1.1. *Simulator networking*

Rapid advances in computer and telecommunications technology during the last decade have led to numerous innovations for supporting training information and instruction delivery systems. Interactive, computer-assisted instruction, high-resolution multimedia delivery systems, distance learning, and use of the Internet as educational media, are but a few of the technology-based techniques that are growing rapidly in sophistication, utility, and acceptance. In addition, driven primarily by military research and development, simulator networking has emerged as a significant new capability for addressing major training issues facing the component military services. One major effort in this area, known as SIMNET, was undertaken partly in response to the requirement for new training methods (Alluisi 1991). SIMNET's networking approach assumed the need to link a multitude of military training 'interfaces' via a common communication format that would allow various land, sea, and aviation components to participate in large-scale exercises. Simulator networking represents a conceptually powerful tool for training. Not only can geographically dispersed personnel and training assets be linked, but also the participants can interact as a team to develop a better understanding and appreciation of one another's capabilities, execution strategies, etc. Typically, it is far too costly to bring actual equipment and personnel assets of geographically dispersed components together for training on a continuing basis. To contain costs and maintain high training levels, additional emphasis will probably be placed on the use of networked simulation in support of distributed training requirements.

As technology is developed to make simulator networking a more generally available and practical training medium, further issues related to the utilization of limited, and, often high-cost simulation assets in networked exercises will need to be addressed. An exemplary problem concerns the dedication of simulators and training devices for network-based training which, for a period of time, makes these capabilities and the involved personnel unavailable to support other organizational training needs. Also, the effectiveness of the networking medium in meeting stated training objectives requires continual evaluation.

The methodology discussed herein was developed to address training requirements in military aviation situations that might be met through the use of networked simulation. In US military aviation, missions vary greatly across the services as well as across the individual aircraft platforms within each of the services. Dedicated training involving simulators and other devices is conducted within each platform area to ensure that operators can perform assigned tasks at acceptable levels under actual operational conditions. At the present time, however, few aircraft platform trainers are networked to allow for inter-platform team training. From a practical standpoint, a methodology is needed that makes possible within- and across-platform training requirement comparisons, and that may be conducted economically within a reasonable time frame. The technique presented here offers components of such a methodology.

2. Methodology development

2.1. *Collective task analysis*

The task and training requirements analysis methodology (TTRAM) was developed to identify collective task training requirements across a variety of US military aviation platforms, and to evaluate the utility of potential training technologies to address these needs. The overall components of the methodology involve identifying existing practice/training gaps, and identifying potential training solutions. Data from these analyses provide a way to identify practice/training support requirements resulting from the lack of availability of training resources (e.g. simulators, training materials, flight hours), and/or infrequent practice. Further, available and emerging training technologies are examined to determine how identified practice/training needs may be satisfied.

A task analysis serves as the basic input to the TTRAM. Unlike traditional task analytic frameworks, however, the analysis used here provides detailed information about team tasks and processes involved in tactical environments. The focus of the analytic approach on team task performance is especially important as it takes into account the extensive requirements for aircrews to communicate and co-ordinate their actions with other aircraft and controlling agencies in carrying out joint missions.

2.2. *TTRAM analytic approach*

The TTRAM utilizes an analytic process that relies upon a series of integrated activities designed to culminate in appropriate training technology recommendations in teamwork intensive tasks. Each component of the methodology is described in detail below, and relevant data collection instruments are provided. The overall process involves: (1) identifying the existence of practice/training gaps; and (2) where gaps exist, specifying potential training solutions. The former activity is supported by a skill decay analysis and a practice analysis. The skill decay analysis component is designed to identify tasks that are susceptible to skill degradation if not adequately practised, and therefore are candidates for additional training support. The practice analysis is intended to identify the amount, frequency, and quality of the existing practice regimen for each identified task.

The generation of potential training solutions is accomplished by performing a training technology analysis that identifies appropriate training technological support to fill identified gaps and/or augment training in critical areas. This analysis also explores the extent to which networked simulation may function to satisfy specific mission rehearsal and training needs. Finally, a set of recommendations is generated that addresses training alternatives (including networked simulation) for training specific skills and performances. Since an important objective of TTRAM is to determine task areas that might benefit from expanded training/practice opportunities, it is important to establish not only the skill decay potential and relative levels of practice support associated with the tasks, but also which tasks require a dynamic exchange of information, high levels of response co-ordination, and precise sequencing of activities with other agencies. Figure 1 provides an overview flowchart of the TTRAM technique. Each component step is discussed below.

2.2.1. *Conduct training analysis*:   The purpose of this initial TTRAM activity is to identify tasks that may require additional practice and/or training support. Two component analyses are performed in order to achieve this goal: (1) a skill decay analysis, and (2) a practice analysis. Both analyses require that characteristics associated with each task be rated by domain experts, knowledgeable about how mission tasks are performed and practised. Task ratings are subsequently transformed into numerical values and summarized into two distinct indices; a skill decay index (which estimates the extent to which a task is prone to degrade if not

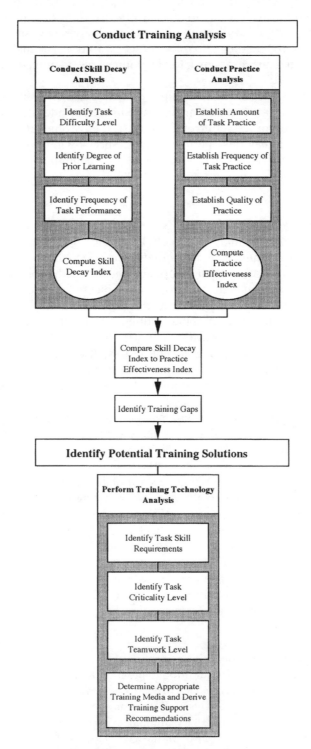

Figure 1. Task and training requirements analysis methodology (TTRAM): process flow diagram.

practised), and a practice effectiveness index (which indicates the level of practice support currently provided for each task). Practice and/or training gaps are identified on a task-by-task basis by assessing the extent to which a task is identified as being prone to decay relative to the practice and training support it receives (i.e. by comparing a task's skill decay index to its practice effectiveness index).

2.2.1.1. *Conduct skill decay analysis*:   Research has indicated that retention of trained skills decreases with time and is influenced by a variety of variables, including length of the retention interval, type of task, and practice or interfering activities performed during the retention interval (Schendel *et al.* 1978). Further, skills that are infrequently used, as well as those that are based primarily on cognitive processes (as opposed to perceptual-motor processes), are known to be particularly susceptible to decay (Childs and Spears 1986). Thus, if many of the skills acquired through training are to be maintained, they must be periodically refreshed (Schendel and Hagman 1982). Factors that contribute to skill decay can be grouped into the following two basic areas: (1) task-related constructs such as the nature of the task, it's difficulty and frequency of task performance; and (2) user-centred factors, including the familiarity and experience of the performers. Information regarding each of these areas is collected and used in the TTRAM as a basis for determining the nature and magnitude of skill/performance loss.

The first step in the TTRAM methodology, termed skill decay analysis, involves determining task skills that are prone to decay if not refreshed or practised adequately. In order to estimate the nature and magnitude of skill loss, information regarding three key factors known to influence skill degradation is collected. The three factors are: task difficulty, degree of prior learning, and frequency of task performance. These specific factors are among the few variables known to influence the rate of forgetting of learned knowledge and skills (Farr 1987).

(1) *Identify task difficulty*: This component addresses the extent to which tasks are inherently difficult to perform, including difficulty in both acquiring and retaining requisite skills. This type of information is important since task difficulty has been shown to influence retention (Lane 1987). Difficulty demands associated with each task are assessed via subject matter expert (SME) interviews, using behaviourally anchored rating scales (BARS), which provide benchmarks and outline the characteristics associated with low, medium, and high difficulty levels (table 1).

(2) *Identify degree of prior learning*: This item addresses the degree of prior learning (proficiency) possessed by the personnel who perform the specific tasks under investigation. This information is important since degree of original learning has been shown to be the single most important ingredient in determining skill retention (Schendel *et al.* 1978, Gardlin and Sitterley 1972). Degree of prior learning, indexed by level of proficiency at the time of training, is also assessed via SME interviews, using BARS. This issue is addressed by establishing the level of training estimated by SME personnel to be required in order to maintain adequate proficiency on each task under investigation. Table 2 outlines characteristics associated with low, medium, and high levels of proficiency.

(3) *Identify frequency of task performance*: This item addresses the length of the skill retention interval, as indicated by the frequency of task performance on

the job. Research has indicated that long periods of inactivity in using skills, or infrequent skill use, are associated with poorer retention; many skills

Table 1. Task difficulty behaviourally anchored rating scale.

Question: How difficult is this task to perform?

Difficulty levels	Associated task characteristics
Low	Virtually no practice is required. Most trained individuals (i.e. 90%) will be able to perform this task with minimal exposure or practice on the operational equipment
	Consists of very few procedural steps, and each step is dependent upon preceding steps
Medium	Individuals can accomplish most of the activity subsequent to baseline instruction. The majority of trained individuals (i.e. 60%) will be able to perform this task with minimal exposure or practice on the operational equipment
	The activity does require moderate practice to sustain competent performance at the desired level of proficiency
	Consists of numerous complex steps
High	Requires extensive instruction and practice to accomplish the activity. Very few trained individuals (i.e. 10%) will be able to perform this task with minimal exposure or practice on the operational equipment
	Consists of a large number of complex steps, and there is little if any dependency among the task steps

Table 2. Degree of prior learning assessment scale.

Question: What level of training is required to maintain an adequate level of proficiency on this task?

Proficiency levels	Associated task characteristics
Low	A high level of training is required to maintain proficiency on this task
	Individual cannot be expected to perform the task without frequent recurrency training
	Individual fails to meet task performance standards without frequent recurrency training
Medium	A moderate level of training is required to maintain proficiency
	Individual can perform the task in the trainer under a restricted set of task conditions; however, needs more practice in the actual job setting under varying task conditions and under supervision
	Individual meets minimum performance standards without frequent recurrency training
High	Minimal training is required to maintain proficiency
	Individual can perform the task completely and accurately without supervision across varying task conditions; has achieved mastery level proficiency
	Individual exceeds performance standards

decay without periodic use (Hagman and Rose 1983). Task frequency judgements are gathered via the assessment scale shown in table 3.

(4) *Compute skills decay index*: A skills decay index is generated on the basis of the foregoing information (i.e. task difficulty, degree of prior learning, and frequency of task performance). This index provides a summary measure intended to indicate which team-oriented tasks are susceptible to decay (i.e. those with significant potential for skill degradation, if not practised). The skill decay index is computed by first translating data relating to each of the three task dimensions into numeric values based on levels assigned during the analysis (e.g. low, medium and high). Scoring algorithms are applied to these aspects as follows:

Task difficulty	1 = low,	2 = medium,	3 = high
Degree of prior learning	1 = high,	2 = medium,	3 = low
Frequency of task performance	1 = high,	2 = medium,	3 = low

The skill decay index is derived by summing across these items to yield a total score, ranging from 3 to 9 points (higher values correspond to increasing levels of potential skill decay). Previous research has indicated that composite scores based on simple linear combinations of two or more ratings are more representative of the process used by raters to form overall judgements than non-linear multiplicative transformations, and are often more reliable than overall judgements (Sanchez and Levine 1989). Thus, estimates of a task's skill decay potential derived using this analytic process should be more valid and reliable than simply asking individuals (however expert they may be in the specific subject matter of interest) to simply form an overall judgement of each task's skill decay potential.

Estimates of each task's skill decay potential may be independently validated (to the extent possible) by collecting additional judgements from training experts. This additional rating information may then be compared against data generated by the skill decay analysis on a task by task basis.

2.2.1.2. *Conduct practice analysis*: Data derived from the skill decay analysis provide estimates of tasks that are prone to decay, but do not address whether these

Table 3. Frequency of task performance assessment scale.

Question: How often is this task performed in the context of your job (across different missions)? Do not factor in time spent training; limit responses to the frequency with which the task is inherently performed as part of the operational setting

Frequency levels	Associated task characteristics
Infrequent	Extremely little time is spent performing this task Task is infrequently performed
Frequent	A moderate amount of time is spent performing this task Task is performed frequently
Very frequent	This task comprises a large amount of time Task is performed very frequently

tasks presently suffer from such deterioration. To address this issue, current practice
and training events are examined. The purpose of the practice/training analysis is to
establish relative levels of task and skill practice provided by current training events
and programmes. During this activity, the existence of practice opportunities
(training and/or practice events) that support each task are identified and further
explored by extracting information relating to the amount, frequency, and quality of
the practice/training. Further, data on existing practice/training requirements
designed to meet the performance objectives associated with each task are collected.
The context in which practice occurs (e.g. actual equipment, whole-task simulator,
part-task simulator, computer-based training, etc.) is also documented. Data
collected during this analysis are then synthesized into an index (referred to as the
'practice effectiveness index'), which indicates the available level of support for each
task.

(1) *Establish amount of practice*: The amount of practice, expressed in terms of
the number of hours/month, associated with each task is elicited from
subject matter expert interviews and documented. Relative numerical
estimates of practice amount may also be represented in terms of high,
medium, and low judgements, where 3 = high, 2 = medium, and 1 = low.
These estimates are established by assessing practice amount for each task
relative to the amounts of practice required for all other tasks in each
mission profile.

(2) *Establish frequency of practice*: The frequency with which tasks are practised
and skills are exercised is also gathered via SME interviews and expressed in
terms of the number of times/month a task is practised. These judgements
may also be expressed in terms of high, medium, and low estimates
(quantified as indicated above). Estimates may be collected for multiple
training cycles where appropriate, to maintain skill proficiency.

(3) *Establish quality of practice*: This item is concerned with the question, 'how
good is the practice?' Factors influencing the quality of practice include the
media used to deliver the practice, the context of the practice, and the
strategies employed in the conduct of practice (Swezey and Llaneras 1992).
Effective team training requires more than just simple practice, it must also
provide performance measurement and feedback, and employ sound
instructional principles (Salas and Cannon-Bowers 1997). This information
is extracted through interviews with training experts who identify, among
other issues, the context in which the practice/training events take place (e.g.
actual equipment, mission-based simulator, part-task simulator, computer-
based training, classroom-based environments).

Table 4 shows four team skill training questions that contain items related
to how well training and practice events support teamwork skills (Morgan *et
al.* 1986). Researchers have emphasized the need to train teamwork skills as
well as to employ measurement systems that are capable of assessing team
processes and outcomes (Swezey and Salas 1992). The four questions shown
in table 4 are designed to capture these dimensions within the context of team
skills training. Again, a three-point scale is utilized where 3 = high,
2 = medium, and 1 = low. These data are collected from SMEs and are
averaged to generate the practice quality index. Data are collected for
multiple training phases of mission activity, if applicable.

Table 4. Team skills training questions.

Extent to which training allows team members to practice co-ordinated activities required by the task (both internal and external to the simulator).	_____
Extent to which training provides practice for improving the effectiveness of communcation among crew members.	_____
Extent to which training incorporates objective measures for evaluating crew performance.	_____
Level of feedback provided by training on how well the aircrew performed as a team.	_____

Directions:
The purpose of these quetions is to assess how well current training and practice exercises support team skill acquisition and maintenance associated with each task. Please indicate the level (H, M or L) at which training activities support task functions (e.g. co-ordination and communication) required in the context of team performance. Enter a High (H), Medium (M), or Low (L) rating for each item.

(4) *Simulator capabilities*: To gain additional perspective on practice quality (although not incorporated into the practice quality index) all relevant training simulators are investigated. Table 5 shows a simulator capabilities and training checklist designed to extract specific information regarding simulator capabilities, as well as how simulators are used to support training and practice events. Merely placing trainees in networks, for example, does not necessarily produce effective team training. Instructional support tools and strategies that enable team members to understand and observe the relevant skill and knowledge components, to practise them, and to receive feedback, are vital to effective training (Salas *et al.* 1995). The checklist includes items that are structured around three unique evaluation categories derived from research conducted by Andrews *et al.* (1992). (These are identified as 'management', 'simulation', and 'instructional technology' functions in table 5.) Together, these items provide a relatively comprehensive picture of a simulator's functional capabilities for training. Items represented in the checklist also address aspects of how the simulator is used to support training, including use of instructional principles for the acquisition, retention, and transfer of skills. Other sources for deriving estimates of practice quality regarding the features and characteristics associated with the practice/training devices are also examined, including technical manuals and documentation.

(5) *Compute practice effectiveness index*: Given that practice/training support currently exists for a task, data are combined relating to the amount, frequency, and quality of practice to yield an estimate of the effectiveness of this support as it relates to skill acquisition, retention, and transfer. This estimate is generated on the basis of the foregoing information (i.e. practice amount, frequency, and quality) by first translating each corresponding data item into a numeric value based on the level of support ($1 = $ low, $2 = $ medium and $3 = $ high). Values for each component of the practice effectiveness index are then added, and may range from 3 to 9 points (higher

Table 5. Simulator capabilities and training checklist.

Tell us about the simulator's capabilities
Check the box if the simulator possesses the following capabilities.

A. *Training management functions*
1. The simulator allows for varying levels of mission/task difficulty.   ☐
2. The simulator supports end-of-session training summaries.   ☐
3. The simulator provides a library of baseline mission scenarios.   ☐
4. The simulator provides a method for placing platforms at specific   ☐
    locations within the exercise boundaries, ready for exercise start.

B. *Training simulation functions*
5. The simulator provides essential sensory cues (motion, visual, auditory).   ☐
6. The simulator provides essential equipment controls.   ☐
7. The simulator provides essential operational or functional characteristics   ☐
    (threat warnings, malfunctions, weapons systems, etc.).
8. The simulator replicates environmental factors found in the operational   ☐
    setting.

C. *Instructional technology functions*
9. The simulator supports objective performance measurement.   ☐
10. The simulator provides post-exercise debrief capabilities (freeze, record,   ☐
    playback, time-stamps, graphic reconstruction of mission events).
11. The simulator allows the entire crew to practice co-ordinated tasks.   ☐
12. Capability exists to practice co-ordinated missions with other aircraft or   ☐
    platforms.   Subtotal _____

Tell us about your simulator-based training
Check the box if training using the simulator incorporates the following features.

1. Training incorporates immediate performance feedback.   ☐
2. Training is distributed across multiple practice sessions.   ☐
3. Training is structured such that difficult task or mission elements are   ☐
    learned independently.
4. Training emphasizes complete MASTERY of mission tasks and skills.   ☐
5. Training provides for frequent and regular testing.   ☐
6. Training conditions increase in similarity to the operational environment   ☐
    as practice/training progresses.
                                                                       Subtotal _____

values correspond to better practice support). Multiple indices may be computed if appropriate, and then averaged to provide a single practice effectiveness index, which is subsequently used to identify practice/training gaps.

2.2.1.3. *Compare skill decay index to practice effectiveness index*: Each task analysed using this methodology has an associated skill decay index and a practice effectiveness index. Both indices range from 3 to 9, with higher skill decay index scores representing greater potential for skill degradation, and higher practice effectiveness scores indicating greater task skill support. As outlined below, the existence of practice/training gaps are identified by comparing the scores on both indices for each mission task.

2.2.1.4. Identify training gaps: Gaps represent areas of task practice/training in which mission task skills are either not addressed, or inadequately addressed by current practice and training regimes. Gaps in current practice and training programmes are identified by contrasting the extent to which task skills are prone to decay against the degree to which existing practice and training regimes address these needs. Gaps are said to exist if skills associated with a task are susceptible to decay, but are not adequately addressed by existing training. Within the context of the present methodology, gaps are identified by comparing the skill decay index (which assesses the skill decay potential of tasks) to the practice effectiveness index (which measures the degree to which existing training adequately supports task skills). Instances where the skill decay index for a given task exceeds the practice effectiveness index, are candidates for additional practice and training support.

The TTRAM also highlights areas where training resources may be better utilized. For instance, tasks with very low skill decay indices and extremely high practice effectiveness indices may represent areas where existing practice and training are currently over-emphasized.

2.2.2. Identify potential training solutions: The purpose of the analytic activities in this section is to specify available instructional technologies, or 'vehicles' (e.g. simulators, computer-based training devices, simulator networking programmes, etc.) which are generally capable of conveying or presenting practice and/or training for mission tasks. In addition, a set of practice/training recommendations (or technology assessments) is generated in order to match alternative training technologies to mission task support requirements.

2.2.2.1. Perform training technology analysis: The purpose of the training technology analysis is to identify available alternative technological practice/training solutions that are appropriate for the unique task performance requirements of specific training situations. This analysis requires that skills underlying each mission task be identified in order to determine appropriate training media.

(1) *Identify task skill requirements*: Research on skill retention suggests that the shape of a specific forgetting curve is influenced by the nature of the task; different types of skills are forgotten at different rates (Farr 1987). In TTRAM, therefore, tasks extracted from a precursor task analysis are categorized in terms of their underlying processes (perceptual, cognitive, psychomotor, etc.) in order to extract information about potential rate of skill loss.

A behavioural classification system (based upon the work of Berliner *et al.* (1964) and of Cannon-Bowers *et al.* (1995) is used to categorize and decompose performance requirements associated with each sub-task. Skill components underlying each sub-task are classified into the skill/knowledge categories listed in table 6. Skill classification categories represented in the taxonomy encompass both individual (codes A to I) and team-oriented components (codes J to P) including decision-making, communication, leadership, and situational awareness (Cannon-Bowers *et al.* 1995). This framework for organizing and classifying tasks and associated performance requirements may be used as additional input to the TTRAM to aid in

Table 6. Task and skill classification system.

Skill knowledge category	Skill code	Description	Select examples
Detecting	A	Concerns that act of becoming alert to the presence of a signal that could be of special interest in the performance of a job or mission	Visually *detecting* the periscope of a snorkeling submarine during daytime operations *Monitoring* or *searching* airspace for enemy aircraft *Scanning* a display
Identifying and recognizing patterns	B	Involves assigning a name to detected patterns based on identifiable characteristics	*Classifying* a sonar target as 'sub' or 'non-sub' Visually *identifying* a flying aircraft as 'friend' or 'enemy' *Locating* the source of a transmission *Interpreting* information
Procedure following	C	This skill concerns carrying out routinized activity, executed as standard operating procedures in some predetermined sequence	*Recalling* the operation and check-out procedures for a piece of equipment (cockpit checklists)
Developing and dealing with concepts	D	This class of events deals with the formation and use of concepts (conceptual knowledge)	*Comprehending* system functions such as complex relationships between system inputs and outputs *Describing* the purpose and function of system components
Recalling bodies of knowledge	E	Encompasses the learning, recognizing, and recalling of verbal information (facts, principles, rules) needed to function in an operational setting	*Recalling* radio frequencies *Recalling* equipment nomenclature
Decision making and problem solving	F	Involves the application of specific decision models useful in solving problems such as diagnosing equipment malfunctions, choosing tactical operations, and planning alternatives	*Assigning* weapons based on threat evaluation *Choosing* frequencies to search plan *Deciding* among combat tactics *Selecting* a diagnostic and troubleshooting strategy *Evaluating* the effectiveness of an approach
Communicating	G	Involves use of standardized message formats to convey information	*Passing* oral information over the communication network *Commanding* and *directing* others *Instructing* aircraft on landing procedures

Table continued

Table 6. continued.

Skill knowledge category	Skill code	Description	Select examples
Complex-continuous motor skill	H	This category is characterized by motor acts involving continuous control skills which require constant input and feedback	*Tracking* a target, or path *Operating* complex equipment *Positioning* an aircraft over a target
Simple-discrete motor skill	I	Involves simple motor activity which has a definite start and end point	*Pressing* a button to arm weapons systems *Setting*-in a radio frequency
Situational awareness	J	Involves the development of shared models of the team's internal and external environment: includes skill in arriving at a common understanding of the situation and applying appropriate task strategies	*Recognizing* a need for action *Identifying* the source and nature of problems *Detecting* and commenting on deviations
Co-ordination	K	The skill to organize team resources, activities, and responses to ensure that tasks are integrated, synchronized, and completed within established temporal constraints	*Co-ordinating* the timing of activities
Mission analysis	L	The skill to evaluate, synthesize, and organize mission-related information	*Formulating* short- and long-range plans *Establishing* mission requirements and constraints
Leadership	M	The skill to direct and co-ordinate the activities of other team members, assess team performance, assign tasks, motivate team members, plan, and organize, and establish a positive atmosphere	*Delegating* tasks *Focusing* attention on critical aspects of the situation *Requesting* mission relevant information
Assertiveness	N	The willingness to actively participate, state and maintain a position, and initiate action	*Making* suggestions *Asking* questions *Confronting* ambiguities
Performance monitoring and feedback	O	Requires team members to provide, seek, and receive task-clarifying feedback; includes the ability to accurately monitor the performance of team-mates, provide constructive feedback regarding errors, and offer advice for improving performance	*Providing* feedback to crew members on their performance *Monitoring* and cross-checking team member activities
Adaptability	P	The skill to use information acquired from the task environment to adjust strategies through the use of compensatory behaviour and reallocation of intra-team resources	*Altering* behaviour to meet situational demands *Helping* and backing-up others when necessary

determining which available or emerging training technologies are candidates for application.

(2) *Identify task criticality level*: Task criticality ratings index the extent to which a task is a critical component of the overall duty/mission. Measures of task criticality have been shown to be a significant component in determining team task importance (Bowers *et al.* 1994) and are therefore included in this analysis. SME judgements using the scale shown in table 7 are used to identify critical team-dependent tasks.

(3) *Identify task teamwork level*: The degree to which a task demands co-ordinated activities and interaction among individuals is assessed via subject matter expert ratings. Key players involved in the execution of tasks, as well as the specific actions and interdependencies among these individuals, are identified either in the pre-established team task analysis or as a part of this analysis. As indicated in table 8, separate estimates of internal and external teamwork are generated. These estimates are used to drive decisions concerning both within- and between- simulator networking needs.

(4) *Determine appropriate training media and derive training support recommendations*: Although a variety of training media may be used in operational settings (e.g. classroom lecture, whole-task simulators, part-task trainers, and computer-based training), reliance on these media, particularly simulators, may vary widely across training situations. Further, existing technologies provide limited capabilities for addressing large scale integrated teamwork training requirements. Table 9 presents a system for classing different types of practice and training media based in part upon the work of Salas and Cannon-Bowers (1997). As illustrated, various forms of instructional technologies exist, ranging from complex electronic devices (simulators, computer-based training devices), to simpler devices such as slides, photographs, and charts. These are identified as information-based, demonstration-based, and practice-based technologies. Unique capabilities

Table 7. Task criticality assessment scale.

Question: How critical is this task to successful mission performance?

Criticality levels	Associated task characteristics
Low	Errors are unlikely to have any negative consequences to overall mission success Task is not a critical/important component of the overall duty/mission Task can be ignored for long periods of time
Medium	Errors or poor performance would have moderate consequences and may jeopardize mission success Task is somewhat critical/important to the overall duty/mission Task requires attention, but does not demand immediate action
High	Errors would most likely have serious consequences, failing to execute this task correctly would lead to mission failure Task is a critical/important component of the overall duty/mission Task requires immediate attention and action

Table 8. Teamwork assessment scale.

Question: What level of teamwork is required in order to perform this task? Assign two ratings, one for internal crew member teamwork, and a second for external teamwork.

Teamwork levels	Associated task characteristics
Low	Task can be accomplished on the basis of individual performance alone; the task can be performed in isolation of other tasks
	Virtually no interaction or co-ordination among team members is required
	Task can be performed in parallel with other team member tasks
Medium	Requires a moderate degree of information exchange about internal/external resources, and some task interdependencies among individuals exist
	Some co-ordination among team members is required if the task is to be successfully completed
	Some sequential dependencies among sub-tasks are required
High	Involves a dynamic exchange of information and resources among team members
	Response co-ordination and sequencing of activities among team members is vital to successful task performance (activities must be synchronized and precisely timed)
	Actions are highly dependent upon the performance of other team members

and features associated with each family of technological training devices make them suited for different instructional applications.

Characteristics specific to simulation-based technologies (i.e. simulators and networked devices), for instance, make them variously appropriate for supporting training management, training simulations, and instructional technology functions (Andrews *et al.* 1992). Training management functions, for example, deal with managing the practice/training environment (such as degree to which the system supports end-of-session training summaries, and generation of exercise scenarios). Training simulation functions refer to the capability to simulate the physical and functional fidelity of real-world environments, while instructional technology functions concern the extent to which the technology supports features known to enhance skill acquisition, retention, and transfer to operational settings (such as the capability to support both objective and subjective performance measurement, and feedback to be presented during various mission phases).

Devices classified as 'networked simulations' expand upon 'simulation' capabilities by providing task and teamwork skill development opportunities among various devices involved in co-ordinated activities. Although both simulated equipment/setting and networked simulations deal with simulation-based technologies, the latter category encompasses a much broader definition of teamwork, i.e. large-scale co-ordination and integration of task activities across diverse training situations.

Teams must learn to co-ordinate their actions so that individual team members can gather, process, integrate, and communicate information in a timely and effective manner. Team training must therefore allow for information exchange, demonstra-

Table 9. Practice/training media classifications.

Types of practice and training technologies	Examples
Practice-based	
Actual equipment/setting	Equipment used in performance of actual real-world mission
Simulated equipment/setting	Whole-task simulators
	Part-task simulators
	Equipment mock-ups
	Role-playing
Networked simulations	Distributed interactive simulation (DIS)
	What-if stimulation system for advanced research and development (WISSARD)
Computer based	Computer-based instruction
	Computer-aided instruction
Demonstration-based	
Video-motion	Videodisc
	Film—Animation
	Videotape
	Television
Information-based	
Projected-still	Slides
	Transparencies
	Picture
Audio	Tapes
	Records
	Compact disc
Printed materials	Diagrams
	Job aids
	Books
Direct classroom instruction	Lecture
	Demonstration

tion of teamwork behaviours, practice, and feedback (Salas and Cannon-Bowers 1997). Although simulations have been used in the past for training small teams, simulator networking allows larger groups of trainees to practise their skills in large-scale interactive simulations (Weaver *et al*. 1995).

In order to assess the utility of networked simulation with respect to other available technological options for providing practice and training support, tasks that involve interdependence and interaction among team members must be identified. The degree to which a task requires high degrees of external teamwork (co-ordination with sources external to the individual task area) and internal teamwork (co-ordination with sources internal to the task area) is documented as part of this process. This approach is similar to that of other researchers (Bowers *et al*. 1993, 1994), where subject matter experts provide information on each task having interdependency among team members. Further, an index of task criticality has been incorporated into this methodology to establish the relative importance of collective tasks in executing the overall mission.

Appropriate training technologies capable of supporting task practice and training are determined by first considering each task's skill decay index as compared to its practice effectiveness index score for each phase of an operation. In addition, important skills underlying task performance, the task's criticality, and internal and external teamwork requirements, may also be considered in deriving potential instructional technologies that can be used to support collective task practice and training needs. Table 9 provides categorized alternatives that may be considered for this purpose.

It is important to note that the recommendations do not include the use of actual operational equipment in support of analysed tasks. Although use of actual operational equipment is an important and undeniably effective way of maintaining skill proficiency, the TTRAM focuses upon *supplemental* ways of augmenting and enhancing skills that are susceptible to decay when adequate levels of practice and training opportunities are lacking. Expanding and/or upgrading existing levels of training support by capitalizing on advances in technology appear to be plausible ways to offset the constraints that limit the use of operational assets in building and sustaining critical mission task skills.

Five major factors are thus considered in establishing training media recommendations. First, what was the relative level of skill decay for the task? Second, what was the relative level of practice effectiveness for the task? Third, was a training gap identified in the comparison of skill decay and practice effectiveness indices? Fourth, was the task rated as critical to the prosecution of the mission? Fifth, was the task determined to require significant levels of external teamwork? The requirement for external teamwork is an effective filter and important discriminator in determining a task's candidacy for training using networked simulation. In particular, even though a task has a high skill decay index, a low practice index, and is judged to be highly critical to a mission, its candidacy for practice via networked simulation is questionable if there are very low external teamwork requirements.

Use of simulation-based practice, particularly networked simulations, appears to be an effective way to augment training that will support many team-based performance skills (Andrews *et al.* 1992). While numerous tasks are currently practised via simulation-based technologies, networked simulation involving wide-spread interaction and co-ordination is not widely available. The development and application of the TTRAM anticipates the introduction of networked simulations as a step in enhanced training. Future work will continue to validate the assumptions underlying the methodology.

3. Summary

In summary, the TTRAM provides a set of analytic tools necessary to systematically address the collective task structures and requirements comprising a broad spectrum of tasks and missions. Experience in using the methodology across numerous US military aircraft platform areas has shown the analytic techniques to effectively discriminate tasks that are prone to skill decay, that are critical to mission success, that require high levels of internal and external teamwork, and that require additional training support. Of equal importance, the TTRAM provides a framework that can be used to evaluate analytic findings from diverse operational and training environments. The resulting reference base should prove valuable in future efforts to establish team training requirements.

Acknowledgements

The authors wish to acknowledge the contributions to this paper of Robert E. Llaneras, Lisa L. Swezey and Frank McCoy. The views expressed are those of the authors and do not reflect the official position of the US Government.

References

ALLUISI, E. A. 1991, The development of technology for collective training: SIMNET, a case history, *Human Factors*, **33**, 343–362.

ANDREWS, D. H., WAAG, W. L. and BELL, H. H. 1992, Training technologies applied to team training: military examples, in R.W. Swezey and E. Salas (eds), *Teams: Their Training and Performance* (Norwood, NJ: Ablex).

BERLINER, D. C., ANGELL, D. and SHEARER, J. 1964, Behaviors, measures, and instruments for performance evaluation in simulated environments. Paper delivered for a symposium and workshop on the Quantification of Human Performance, Albuquerque, New Mexico, August.

BOWERS, C. A., BAKER, P. and SALAS, E. 1994, Measuring the importance of teamwork: the reliability and validity of job/task analysis indices for team-training designs, *Military Psychology*, **6**, 205–214.

BOWERS, C. A., MORGAN, B. B., JR, SALAS, E. and PRINCE, C. 1993, Assessment of coordination demand for aircrew coordination training, *Military Psychology*, **5**, 95–112.

CANNON-BOWERS, J. A., TANNENBAUM, S. I., SALAS, E. and VOLPE, C. E. 1995, Defining team competencies: implications for training requirements and strategies, in R. Guzzo and E. Salas (eds), *Team Effectiveness and Decision Making in Organizations* (San Francisco: Jossey-Bass).

CHILDS, J. M. and SPEARS, W. D. 1986, Flight-skill decay and recurrent training, *Perceptual and Motor Skills*, **62**, 235–242.

DYER, J. L. 1984, Team research and team training: a state-of-the-art review, in F.A. Muckler (ed.), *Human Factors Review* (Santa Monica, CA: Human Factors Society), 285–323.

FARR, M. J. 1987, *The Long-term Retention of Knowledge and Skills: A Cognitive and Instructional Perspective* (Berlin: Springer-Verlag).

FORSYTH, D. R. 1983, *An Introduction to Group Dynamics* (Monterey, CA: Brooks/Cole).

GARDLIN, G. R. and SITTERLEY, T. E. 1972, Degradation of learned skills: a review and annotated bibliography. Report D180-15080-1, Boeing, Seattle, WA.

HAGMAN J. D. and ROSE, A. M. 1983, Retention of military tasks: a review, *Human Factors*, **25**, 199–213.

LANE, N. E. 1987, *Skill Acquisition Rates and Patterns: Issues and Training Implications* (Berlin: Springer-Verlag).

LEVINE, E. L., PENNER, L. A., BRANNICK, M. T., COOVERT, M. D. and LLOBET, J. M. 1988, Analysis of job/task analysis methodologies for team training design. Contract no. DAAL03-86-D-0001 Naval Training Systems Center, Orlando, FL.

MCRAE, A. V. 1966, Interaction content and team effectiveness, HUMRRO-TR-66-10, AD-637311. Human Resources Research Office, George Washington University, Alexandria, VA.

MORGAN, B. B., JR., GLICKMAN, A. S., WOODWARD, E. A., BLAIWES, A. S. and SALAS, E. 1986, Measurement of team behaviors in a Navy environment, Technical Report no. NTSC TR-86-014, Naval Training Systems Center, Orlando, FL.

OBRIEN, G. E. and OWENS, A. G. 1969, Effects of organizational structure on correlations between abilities and group productivity, *Journal of Applied Psychology*, **53**, 525–530.

SALAS, E. and CANNON-BOWERS, J. A. 1997, Methods, tools, and strategies for team training, in M.A. Quinones and A. Ehrenstein (eds), *Training for 21st Century Technology: Applications of Psychological Research* (Washington, DC: APA Press), 249–280.

SALAS, E., BOWERS, C. A. and CANNON-BOWERS, J. A. 1995, Military team research: ten years of progress, *Military Psychology*, **7**, 55–76.

SANCHEZ, J. I. and LEVINE, E. L. 1989, Determining important tasks within jobs: a policy-capturing approach, *Journal of Applied Psychology*, **74**, 336–342.

SCHENDEL, J. D. and HAGMAN, J. D. 1982, On sustaining procedural skills over a prolonged retention interval, *Journal of Applied Psychology*, **67**, 605–610.

SCHENDEL, J. D., SHIELDS, J. L. and KATZ, M. S. 1978, Retention of motor skills: review, ARI Technical Paper 313, US Army Research Institute for the Behavioral and Social Sciences, Alexandria, VA.

SHAW, M. E. 1976, *Group Dynamics: The Psychology of Small Group Behavior* (New York: McGraw-Hill).

STREUFERT, S. and SWEZEY, R. W. 1986, *Complexity, Managers and Organizations* (New York: Academic Press).

SWEZEY, R. W. and LLANERAS, R. E. 1992, Validation of an aid for selection of instructional media and strategies, *Perceptual and Motor Skills*, **74**, 35.

SWEZEY, R. W. and LLANERAS, R. E. 1997, Models in training and instruction, in G. Salvendy (ed.), *Handbook of Human Factors*, 2nd edn. (New York: Wiley).

SWEZEY, R. W. and SALAS, E. 1992, Guidelines for designing team training systems, in R.W. Swezey and E. Salas (eds), *Teams: Their Training and Performance* (Norwood, NJ: Ablex).

SWEZEY, R. W., MELTZER, A. L. and SALAS, E. 1994, Issues involved in motivating teams, in H.F. O'Neil, Jr. and M. Drillings, (eds), *Motivation: Theory and Research* (Hillsdale, NJ: Lawrence Erlbaum), 141–170.

WEAVER, J. L., BOWERS, C. A., SALAS, E. and CANNON-BOWERS, J. A. 1995, Networked simulations: new paradigms for team performance research, *Behavior Research Methods, Instruments, and Computers*, **27**, 12–24.

CHAPTER 10

Cognitive task analyses for decision centred design and training

DAVID O'HARE†, MARK WIGGINS‡, ANTHONY WILLIAMS† and WILLIAM WONG§

†Department of Psychology, University of Otago, Box 56, Dunedin, New Zealand

‡Department of Aviation Studies, The University of Western Sydney, Bankstown Airport, NSW 2200, Australia

§Department of Information Science, University of Otago, Box 56, Dunedin, New Zealand

Keywords: Cognitive task analysis; Critical decision method; Recognition-primed decision making; Training; Systems design.

This paper presents three case studies of Cognitive Task Analysis (CTA) for defining systems design and training requirements. The approach taken involves a modification of the critical decision method of Klein *et al.* The authors utilized the revised CDM to obtain information from expert white-water rafting guides, general aviation pilots, and emergency ambulance dispatchers. The information obtained was used to develop multi-media tools for training rafting guides and general aviation pilots, and to redesign the VDU display requirements for the ambulance dispatchers. The examples demonstrate the utility of an approach to CTA that is closely based on relevant theory, and provides guidance to practitioners wishing to apply CTA techniques.

1. Introduction

Ergonomics and human factors are concerned with the overall improvement of system effectiveness by two complementary processes: the design of displays, processes and procedures and the selection and training of personnel. The two are complementary, as Roscoe (1980) observes since 'the former serves to reduce the need for the latter; the latter completes the job left undone by the former' (Roscoe 1980: 3). The aim of the present paper is to demonstrate how cognitive task analysis (CTA) can be used to support both processes. The authors illustrate this with examples from recent studies of emergency ambulance dispatchers, aircraft pilots, and white-water rafting guides.

Cognitive task analysis may be defined as an approach that 'determines the mental processes and skills required to perform a task at high proficiency levels and the changes that occur as skills develop' (Redding 1992: 3). CTA is appropriate for tasks that are cognitively complex (requiring an extensive knowledge base, complex inferences and judgement) and which take place in a 'complex, dynamic, uncertain, real-time environment' (Gordon 1995: 528). CTA may be carried out using a wide variety of techniques. The most commonly used techniques, according to Redding and Seamster (1994), include cognitive interviewing, protocol analysis, scaling techniques such as multidimensional scaling (MDS), neural network modelling, computer simulations, and error analysis.

These various CTA techniques can be divided into two distinct subgroups (Roth and Mumaw 1995, Vicente 1995). One approach focuses on identifying the inherent constraints in the application domain. The analysis is based on a rational decomposition of the functional structure of the application domain. A major advantage of this approach is that it can be used on prototype systems not yet in operation. This form of CTA is usually associated with the design and development of the operator interface. The second approach is based on an analysis of actual user activity in already functioning systems and tends to be associated with the design and development of operator training programmes.

The first group of techniques tend to draw their theoretical support from the functional approach of Rasmussen (1986), and increasingly, the 'ecological' approach of Gibson (1979). This has led to the development of a coherent and integrated approach to problems of interface design in complex systems. The second group of techniques relies on the traditional empirical task analysis approach (cf. Drury *et al.* 1987), which is best suited to highly proceduralized tasks and has been less successful in dealing with tasks with a high level of variability in operator strategies (Vicente 1995). The successful analysis of such cognitive skills requires an appropriate terminology and explicit theoretical basis. One theoretical approach and its associated CTA technique that has gained widespread attention is the Recognition-Primed Decision (RPD) model of Klein (1989) and the Critical Decision Method (CDM) form of CTA described by Klein *et al.* (1989).

The recognition-primed decision model (Klein 1989) emphasizes the situational assessment component of decision making. In traditional models, the emphasis has been on selecting a course of action from a variety of alternatives. In the simplest version of the RPD model the decision maker recognizes the situation as familiar. This recognition incorporates the appropriate goals, cues, expectancies and courses of action for that event. Familiarity may come from previous experience or the clear cut nature of the event (e.g. a 'textbook case'). In more complex cases, the situational assessment phase involves seeking further information until an acceptable situational assessment is reached. A set of goals, cues, expectancies and courses of action are generated by this assessment. The decision maker may conduct a 'mental simulation' of the most plausible course of action to verify its suitability for the circumstances. Even in these more complex cases, the decision maker is more concerned with situational assessment than with weighing the merits of multiple options.

The CTA technique developed by Klein and colleagues to study decision-making in naturalistic environments is known as the Critical Decision Method (CDM). 'The CDM is a retrospective interview strategy that applies a set of cognitive probes to actual nonroutine incidents that required expert judgment or decision making' (Klein *et al.* 1989: 464). The technique was designed to meet three criteria: (1) to address the basis of expert decision-making performance as described by the RPD model; (2) to be applicable under field conditions; and (3) to 'have applied value in terms of training, system design, or development' (Klein *et al.* 1989: 464).

The purpose of this paper is to present the results of the application of a revised form of the CDM approach to CTA in three different domains—aviation, white-water rafting, and emergency control. The aim is to demonstrate the benefits of a theoretically driven CTA in supporting the twin goals of enhanced system effectiveness through design and through training. The authors suggest modifications and extensions to the CDM that will enhance its usefulness in subsequent applications. They then outline possible methods for representing the information obtained from the CTA that bridge the gap between CTA and system (display or training) design. Finally, they briefly outline the details of two multi-media training systems developed from applications of the CDM form of CTA.

2. Method

The CDM as developed by Klein *et al.* (1989) is a descendent of the critical incident technique developed by Flanagan (1954). In common with Flanagan's technique, the CDM explores decision-making in previously experienced non-routine cases. Participants are requested to bring to mind an incident in which their expertise made a difference to the outcome. This incident is recalled in detail and a time-line constructed to establish the precise sequence of events (see study 3 for a useful enhancement to this technique). These actual incidents can themselves be useful as training scenarios or in constructing realistic scenarios for prototype testing.

The unique contribution of the CDM is to examine the critical incidents in more detail using a set of cognitive probes developed from the RPD model. An important feature is the emphasis on 'perceptually based cues that are difficult or unnatural for people to articulate' (Klein *et al.* 1989: 465). These perceptual cues form the basis of the pattern-matching aspect of the RPD model. To the extent that the RPD model of complex skills is valid, then the CDM approach should provide uniquely valuable information to support systems improvement through training and/or design.

The original CDM probes as described by Klein *et al.* (1989) were revised and expanded to obtain additional information on the perceptual and cognitive structures and processes that appear to mediate expertise (Ericsson and Smith 1991). In particular, the revised probes were designed to elaborate on the processes of maintaining situational awareness and planning that previous research has shown to be 'the primary skills distinguishing expert performance' (Redding and Seamster 1994: 205). The revised CDM probes are shown in the appendix.

3. Study 1

White-water rafting is characterized by dynamic and shifting environments where constant risks to one's well-being exist (e.g. risk of exposure, drowning, striking objects). Commercial operators take paying clients, who frequently have little or no rafting experience, on difficult sections of water where the rapids can be up to grade IV or V (VI being the highest and considered impassable even by experienced kayakers). The guide's job is to take up to eight clients on each raft trip, where control over the raft's direction and the rapid decisions he or she makes determine the safety of the trip. Expert raft guides often report encountering problematic situations where urgent decisions are required in a matter of seconds.

3.1. *Participants*

Interviews were conducted with 17 raft guides (15 males, two females) with ages ranging from 17 to 38 years. Five of the guides were classed as expert guides with an estimated average of 2900 commercial rafting trips (minimum: 500). Five guides were categorized as trip leader with an estimated average of 340 commercial raft trips (minimum: 150), and seven guides were categorized as novice with an estimated average of 45 rafting trips (minimum: 25). Participants were selected from four white-water rafting companies, three in Queenstown, and one adventure training company in Nelson, New Zealand.

The raft guides were asked to describe, in as much detail as possible, an incident in which they had to make a critical decision (or series of critical decisions) during a raft trip. It was emphasized that the incidents reported should be in some manner unusual and difficult (i.e. where the best choice was not clear cut) in which the guide felt that their expertise and experience made a critical difference to the outcome. The events of each incident were organized onto a timeline reflecting as accurately as possible the actual sequence and timing of events. Guide decisions were then identified and probed using the revised CDM (See appendix).

After each interview, timelines were matched against tape recorded accounts to ensure the accurate evaluation of each decision event, and decision points were summarized into situation assessment records (Klein *et al.* 1989). Each situation assessment record (SAR) is a breakdown of: the primary cues in the environment being sought or identified by the guides during situation assessment; the expectancies generated by these cues; the consequent goals of the guides; and their selected course of action resulting from this knowledge. Following each SAR is a summary of each guide's decision strategy.

3.2. *Results and discussion*

Seventeen critical incidents were collected during interviews, each containing between 1 and 5 observable decision points, giving a total of 52 decision points. All of the incidents reported were in some way unique or non-routine and the majority had occurred within the 6 months prior to the interview. Nearly all the reported events involved the running of rapids. The most common situation involved retrieving one or more clients from the water. Weather, water levels and consequent river typography could change markedly in under an hour. Additionally, raft direction, crew positioning and river routes could fluctuate severely in a matter of seconds in some cases. Consequently, it appears that river guides are constantly bombarded by a large amount of both relevant and irrelevant information that appears to cause distraction, or a narrowing of the attentional focus, of novice and some trip leader guides. Bystanders on the bank shouting instructions to those in the water was one example of distracting information. While none of the experts reported being affected by such distractions, five of the novice guides and two trip leader guides reported that the additional, or irrelevant, information tended to act as a task distractor.

One of the unique factors about this decision environment is that decisions are required with little or no time for conscious deliberation. The majority (88%) of on-river decisions were judged to be made in less than 5 s, with two decisions (both scouting situations) taking 1 min and 8 min, respectively. On every occasion guides reported having no time to spare when they made their respective decisions.

3.3. *Decision type*

Consistent with the RPD model, expert raft guides tended to consider no more than two action options when making a pressurized decision and described their decision-making for non-routine situations as being highly intuitive in nature. Experts reported considering their options at the time in a singular or serial manner (i.e. experts tended to consider a single course of action and then execute it). One instance was recorded where options were first assessed through mental simulation and then discarded when problems could be foreseen as resulting from such an action.

In comparison, trip leader guides reported a mixture of both intuitive ($n = 2$) and analytical ($n = 3$) modes of decision-making. That is, two of the trip leaders only generated one course of action and went with it (intuitive), and three trip leaders considered up to five action options concurrently until the most appropriate action became apparent (analytical). This analytical decision strategy may be more time-consuming than intuitive decision-making styles as more effort is required to generate and evaluate a higher number of action options. On average trip leaders reported 'extremely high' time pressure during their decision-making. This time pressure combined with the decision strategy of generating and evaluating action options in this more analytical manner may also explain why the estimated mental workload associated with decisions was rated as 'very high' by this group.

One interesting finding was that all of the novice guides made pressurized decisions on the river in a rapid manner without generating a list of action options and concurrently evaluating them. Although at first glance the decision-making processes of novice guides appears similar to that of expert guides, it could be hypothesized that their decision process (i.e. single option generation) is the result of an inability to generate potential action options perhaps resulting from a lack of experience in the operational environment. Such an inability to construct plausible action options may explain why many subjects in this group reported sensations of panic, feelings of helplessness, and decision uncertainty.

Based on these findings it appears that, as predicted by the RPD model, expert guides are able to retrieve the most appropriate option initially without engaging in multi-option comparisons. Trip leader guides tend to employ a mixture of both analytical and intuitive decision styles, whereas novice guides tend to act in accordance with their initially generated course of action. Ramifications for the rafting industry derive from the finding that trip leaders are likely to generate a more thorough list of action options than novices or experts, yet this decision strategy seems to be associated with greater tension, time pressure, and stress than either novices or experts, and greater mental workload than novices. It was also found that trip leaders may experience considerable doubt and uncertainty over selected courses of action and this may also contribute to slower decision-making times on the river. Novice guides are also likely to be slower in identifying appropriate courses of action than experts in pressurized situations, and may hesitate (due to decision uncertainty, panic, feelings of helplessness), before carrying out a selected action. These findings reflect those of Kersholt (1994) who proposed that individuals do not react optimally in dealing with uncertainty and high time pressure when completing a complex decision task.

In essence, as predicted by the RPD model, expert guides appear to use environmental cues to select a course of action by matching the current situation to a similar situation they have encountered in the past, or by matching it to a situation prototype in their memory. Unfortunately it was not established whether the expert guides were drawing on direct analogies of previous experiences to direct their decision-making, or whether they were using a prototype matching approach where incidents are matched to a general prototype or composite schema in memory.

As part of the revised CDM protocol, participants were asked if, during the situation being recalled, they were reminded of previous experiences in which they had made similar or dissimilar decisions. Most of the guides had difficulty in answering this question. Two experts recalled exact instances of a similar nature at the time of the situation, one was unsure, and two expert guides reported the use of a prototype match. For example, one of the expert guides instructed all the rafts to pull out and portage round a rapid that he had had an unpleasant experience in during similar river conditions previously. Two trip leader guides also reported the use of analogical reasoning, however no such use was reported amongst novices. Only one expert guide reported that he did not draw on previous experiences while making his decisions. However, his decisions were still made rapidly (all in less than 5 s), and his options generated serially. Thus it appears that the interview questioning failed to adequately differentiate between the recall of specific incidents for use in problem solving and the use of a generalized or prototype match to aid decision-making.

Klein and Calderwood (1988) suggest that the recall of specific analogues or cases will decline with increasing experience in a domain. In reviewing their CDM findings from over 400 decision points, only 33 direct analogues were identified. They also suggest that the CDM methodology, where the probed events may have occurred some time previously, is itself responsible for the low rate of reported analogue use. The authors would add to this the observation that the CDM question probe does not distinguish between recall of specific analogues and recall of general prototypes. An important issue for research on expert decision making is the development of expert judgement of familiarity or typicality. How are individual cases aggregated? Klein and Hoffman (1993) suggest that experts have amalgamated individual cases together to yield prototypical examples. However, anecdotal evidence would suggest that experts also retain an extensive repertoire of individual cases ('war stories'). Is expert knowledge represented and stored in different ways, or is the retrieval process different?

These results have other theoretical implications for the RPD model itself. The findings suggest that the model fails to adequately describe the situational understanding that is gathered before most problematic situations are encountered. In general, non-routine rafting situations can place guides under a substantial amount of physical, mental, and emotional stress. The most experienced guides reported the continual use of mental simulation to prepare themselves for the river ahead currently out of sight. McLennan and Omodei (1996) have labelled this process as 'pre-priming' and provide evidence for the importance of pre-priming in sports refereeing and firefighting.

3.4. *Implications for guide training*

The findings of the present study have clear implications for the training of individuals at the novice and trip leader level of ability in commercial white-water rafting. Incidents of a non-routine nature involving high risk require rapid decision-making (i.e. in a few seconds), commitment to a course of action, and preferably action selection based on that which has been used successfully in the past. Guide training needs to be specifically tailored to enable guides to recognize pertinent cues in the environment that may be indicative of potential danger while avoiding distracting and irrelevant information.

One possibility is to provide instruction on mental simulation to visualize downriver hazards and to select appropriate lines through rapids. It is clear that mental practice can enhance performance (Driskell *et al.* 1994). Mental simulation involves the visualization of the outcomes of a course of action rather than a mental rehearsal of the actions themselves. Nevertheless, since it was shown that mental rehearsal was particularly effective as the cognitive requirements of the task increased (Driskell *et al.* 1994), it is likely that mental simulation will be a particularly profitable strategy. Another possibility would be to enhance perceptual learning directly through the development of a multi-media package demonstrating the cue recognition of expert river guides. The next section outlines the development of a prototype tool.

3.5. *RAPID: a multi-media tool for perception and judgement training*

An analysis of the cue identification information from the CDM suggested that expert raft guides may take into account a wide range of cues in determining their course of action. One set of cues pertains to the river conditions and the other to the activities of the raft crew and the position of the raft itself. The authors have developed a prototype multi-media tool that can be used to further investigate differences in cue perception and judgement, and can be easily adapted to provide training in cue recognition and judgement. The programme was developed in Oracle Media Objects^{TM} a Hyper-card^{TM}-like authoring system for Macintosh and PC computers [The RAPID multi-media programme was written by Rodney Tamblyn of the University of Otago Computing Services Centre]. RAPID has been tested on Macintosh, Windows 3.1 and Windows95 operating systems. The task is based on viewing brief video segments showing a variety of rafts traversing the same Grade V rapid (figure 1).

Participants are required to stop the video as soon as they have enough information to predict the course that the raft will take through the rapid. They then indicate their prediction on a screen as shown in figure 2.

They are also asked to predict whether the raft will negotiate the rapid successfully. Preliminary trials (Williams 1997) indicate that the task is extremely demanding even for expert guides. A wide range of individual differences are apparent, which may reflect fundamental differences in ability to manage the demands of complex divided-attention tasks (O'Hare 1977, Roscoe 1997). This prediction task can easily be modified to a training tool by providing information about the relevant cues revealed from the CDM analysis. Participants would be able

Figure 1.   The main screen from the RAPID multi-media tool for expert rafting judgement. The participant's task is to pause the video as soon as they feel able to make a prediction as to the course the raft will take on entering the rapid. The vertical bar on the left hand side of the screen indicates time remaining.

Please click on a button to select the route which best approximates the route the raft will take (an animation will illustrate the route). Click "OK" to confirm your selection.

OK

Figure 2. The course prediction screen from the RAPID multi-media tool for expert rafting judgement. By 'clicking' on the points on the right-hand side of the screen, participants can view a projected course (indicated by a moving orange circle) and select the course that they feel was the most likely one for the raft to have taken.

to 'click' on any part of the video to reveal a description of the relevant cues. This approach has been more fully developed as the first part of a multi-media training package for aircraft pilot decision making skills as outlined in the second study below.

4. Study 2

CTA techniques have recently been applied to air traffic control tasks (Schlager *et al.* 1990, Harwood *et al.* 1991, Seamster *et al.* 1993) and to fighter pilot performance (Thordsen *et al.* 1993). However, the majority of pilots are neither trained nor licensed to professional levels and conduct their flying activity under 'visual flight rules' (VFR), which requires them to remain clear of cloud and in sight of the ground at all times. Constantly changing weather conditions form a significant hazard for these pilots and the risks of VFR flying are consequently much greater than for other forms of flying (O'Hare 1990). The authors have begun to examine the skills required for successful VFR flying (Wiggins and O'Hare 1995) with the goal of developing a training programme that will accelerate the development of key elements of VFR decision-making and lead to subsequent improvements in the safety record in this area of aviation.

Traditional instructional materials for pilots tend to concentrate on developing an understanding of weather systems organized around concepts such as frontal systems, pressure, humidity, etc. In order to fly safely in various weather conditions, however, it is necessary to make critical perceptual differentiations not covered by the textbooks. A similar situation with respect to critical care nursing was reported by

Crandall and Getchell-Reiter (1993) whose use of the CDM revealed a number of perceptual cues used by experienced nurses which were not reported in textbooks.

4.1. *Participants*

The participants were 10 experienced general aviation (i.e. non-airline) pilots with an average of over 10 000 flight hours total experience. Since a private pilot licence can be obtained with a minimum of 50 flight hours and a commercial licence after 150 flight hours, this represents a considerable level of experience. Participants averaged over 7500 hours of cross-country flying experience involving over 500 encounters with bad weather. Interviews were carried out individually, and recorded on audio tape. Participants were asked to recall an in-flight situation requiring an unusual and difficult decision with regard to the weather. Two specific weather situations were probed: (a) slowly deteriorating conditions; and (b) thunderstorm activity. The decisions made in these situations were then probed using the revised CDM protocol as shown in the appendix.

4.2. *Results and discussion*

Critical cue inventories (CCIs) (Klein *et al.* 1989), were constructed for each of the two meteorological phenomena. Frequently mentioned cues included cloud-base, cloud-type, height above terrain, horizon, and cloud-colouring. The CDM results revealed a high degree of consistency between experts regarding the critical cues for action in a variety of weather-related decisions. Critical cues are consistently associated with a particular condition such as haze, thunderstorms, frontal conditions, etc. This notion of consistency is, as Fisk and Gallini (1989) observe, pivotal to the notion that practice can lead to fast, accurate, and relatively effortless performance. This kind of automatic processing can only develop where 'invariant rules, invariant components of processing, or invariant sequences of information-processing components are used in successful task performance' (Fisk and Gallini 1989: 454).

4.3. *Implications for training*

The first implication for training is that novices should be trained to develop the perceptual differentiations used by the expert. However learning these cues is one thing; applying that knowledge in different situations is another. In order to develop a procedural skill it is necessary for practice to take place under conditions where certain cues are consistently associated with certain responses. This is what Anderson (1987) refers to as the 'compilation' stage of skill acquisition. The authors have implemented these ideas in a prototype weather-related decision-making tutoring system. Brief details of the system are described below. In order to develop the tutoring system it was necessary to find a technique to structure and integrate the outcomes of the CTA. This stage of development is often left to the judgement of the designer as to how to incorporate the findings of a CTA. Gordon *et al.* (1993) have developed a technique known as Conceptual Graph Analysis (CGA), which can be used to facilitate this process.

4.4. *Conceptual graph analysis*

Conceptual Graph Analysis (CGA) is a technique through which the data arising from cognitive task analytic techniques can be modelled and integrated. Originally developed within the computer science domain, CGA has been adapted to facilitate the representation of cognitive structures amongst subject-matter experts (Gordon *et*

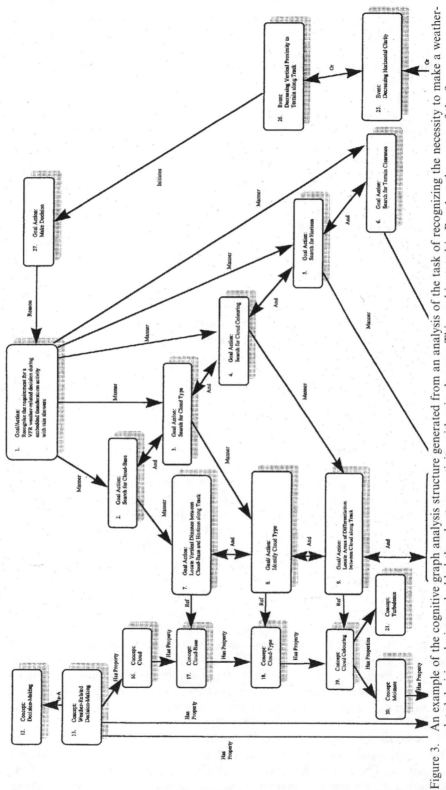

Figure 3. An example of the cognitive graph analysis structure generated from an analysis of the task of recognizing the necessity to make a weather-related decision during embedded thunderstorm activity with rain showers. This goal is stated in Box 1 at the centre of the figure.

al. 1993). Conceptual graphs are designed to provide a diagrammatic structure for the representation of the cognitive elements associated with a particular domain. Gordon *et al.* (1993) refer to cognitive elements as nodes, and these are linked according to a series of rules that define their relationship. There are two categories of node: concepts and statements.

The information within conceptual nodes is similar to declarative knowledge and defines that knowledge of a concept that is required to initiate and/or respond to a particular statement. For example, an appropriate response to a deterioration in the height of cloud above ground is dependent upon an understanding of the concepts of 'cloud' and 'height above the terrain'. Knowledge of these concepts enables the recognition of the situation, and allows an appropriate response to be generated.

Conceptual nodes are typically depicted hierarchically, with the overall concept at the head of the diagram (figure 3). These concepts are linked directly to statement nodes, the information within which can loosely be equated to procedural knowledge as defined by Anderson (1987). Unlike conceptual nodes, statement nodes are divided into one of five categories: goals, goal/action statements, style, events, and states (Gordon *et al.* 1993).

Goal nodes designate desired outcomes, and generally direct the nature and structure of subordinate goals, events and states. In order to minimize the size of conceptual graphs, goal nodes are often associated with action statements to form single goal/action nodes. From the perspective of CGA, the distinction between goal nodes and goal/action nodes appears to rest more upon the principle of parsimony, than any particular theoretical perspective. Nevertheless, the relationship between a goal state and an action state could be regarded as compatible with production system theory, in which condition statements ('If') and action statements ('Then') provide a description of the procedures required for task performance. It might therefore be argued that goal/action statements represent one aspect of the procedural knowledge that is required for effective and efficient task-oriented performance according to production system theory.

The relationship between nodes is specified according to a series of arcs, which are designed to describe both the direction and the nature of the relationship. The main advantage associated with CGA is this specification of the relationships that can occur between nodes. This provides a clear framework within which to represent data, and ensures a level of reliability for the interpretation of data sets. Moreover, these guidelines are based upon a series of theoretical principles devised by Graesser and colleagues (Graesser and Clark 1985), and applied and tested empirically by Gordon *et al.* (1993) within the field of instructional design.

The development of the CGA followed the guidelines suggested by Gordon *et al.* (1993). Conceptual information (e.g. the cues identified above) were listed on the left of the figure with corresponding goal/action statements to the right. The expert pilots were then asked to respond to a series of probes such as 'What happens after (event X)?' or 'Why do you (goal/action Y)?'. Although this was a time-consuming process, it yielded a relatively coherent model that clearly represented both the conceptual (declarative) and goal/action (procedural) components of expert knowledge.

4.5. *Pilot tutoring system for weather-related decision making*
This multi-media system was developed using the Authorware™ system for PC computers. The system has been developed as part of a US Federal Aviation Administration sponsored project on general aviation pilot decision-making under

the direction of Dr Richard Jensen at The Ohio State University (Jensen, 1995). The CGA shown in figure 3 was one example used to structure the material obtained from the CDM/CTA for the development of the instructional software. The first part of the program is concerned with learning to recognize the cues involved. Figure 4 illustrates one of the earlier screens in which participants are able to view in-flight video clips and text information to help them to identify the critical perceptual cues in a given situation. This essentially develops declarative knowledge about flight in these conditions.

Participants are then invited to compare sequences of full-motion video segments to determine where there have been significant changes in the critical cues. By inviting participants to make their own comparisons and receive feedback the programme is helping to develop procedural skill at this complex task. An example of the cue comparison screen is shown in figure 5. This approach is similar to the 'contrast set' approach described by Bransford *et al.* (1989), which requires students to actively identify the features that discriminate between various cases. This is in contrast to traditional approaches, which simply provide students with lists of the characteristics of different cases.

The software is still under development, but responses from flight instructors and others to the prototype system has been extremely positive. After further software development it is planned to conduct several evaluation studies of the acceptability and effectiveness of the instructional system. In contrast to many other 'computer-

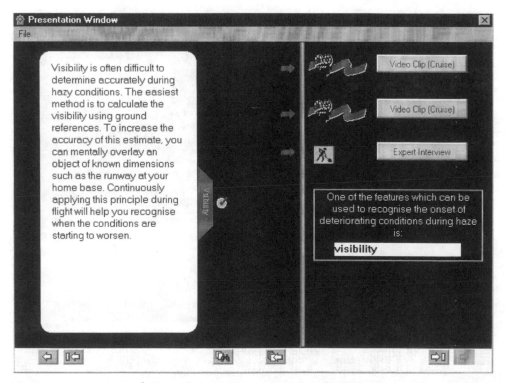

Figure 4. A screen from the cue recognition part of the pilot tutoring system programme. Clicking on different parts of the full-motion video reveals the text box on the right-hand side of the screen, which provides feedback about the importance of that particular cue.

based training' systems in aviation, the focus of this system is on developing procedural skill ('knowing how') rather than declarative knowledge ('knowing what'). While 'Most of the current uses of computer-based instruction are still based on the programmed instructional models developed right before, during, and after World War II' (Brock 1997: 579), the present system is firmly anchored in cognitive science principles of skill acquisition and has been generated from information gathered by CTA and represented and integrated by CGA. The authors have been able to follow a similarly structured approach to a somewhat different question: how to design an effective information display system for emergency ambulance dispatchers?

5. Study 3

Naturalistic decision-making (NDM) environments have been characterized (Orasanu and Connolly 1993) as involving dynamic conditions, time pressure, high-stakes, uncertain information, and multiple players. Examples of NDM environments include air traffic control, nuclear power plant control rooms, military command and control centres, and emergency co-ordination centres. The emergency ambulance control centre has to screen incoming calls, diagnose the nature and severity of the problem, determine available resources and allocate appropriate resources to the problem while maintaining adequate resources for future problems. A large urban centre such as the Sydney Ambulance Coordination Centre is

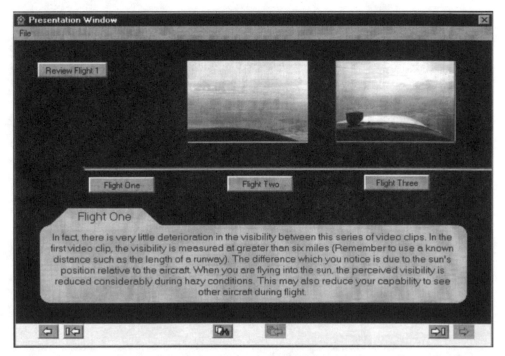

Figure 5.   In this screen from the pilot tutoring system programme, participants compare video clips taken from the earlier and later parts of a single flight in order to determine where there have been significant changes in the cue named at the top of the screen (in this case 'terrain'). Responses are entered and feedback provided in the text box in the lower part of the screen.

responsible for a population of around 4 million spread over an area of 14 000 km². The available resources include 2 helicopters and 130 ambulances spread over 130 stations in the region. In addition to routine jobs such as patient transfers, this centre receives approximately 295 000 emergency calls annually, or around 800 per day.

At the time of the study, the entire process at the Sydney Ambulance Coordination Centre was conducted manually with information transferred on colour-coded 'job slips', which were circulated around the centre on conveyor belts. There would normally be four 'call takers' and five 'dispatchers' working the system at any one time. In addition there would be three supervisory positions. The process is described in more detail in Wong (1996). The aim of the present study was to determine the display requirements for a computer-based system to replace the manual system in current use. A cautionary note is sounded by a previous attempt to computerize a manual ambulance dispatch system. The new London ambulance computer-aided dispatch project (LASCAD) had ended in dismal failure with a night of chaos on 26 October 1992, allegedly resulting in 20–30 deaths due to the late arrival of ambulances (Beynon-Davies 1995).

5.1. *Participants*

Five dispatchers from the Sydney Ambulance Coordination Centre with between 5 and 9 years experience were interviewed for this study. Participants were asked to think about a particularly unusual and challenging incident that they had been involved in. When the initial details of the incident had been produced, the

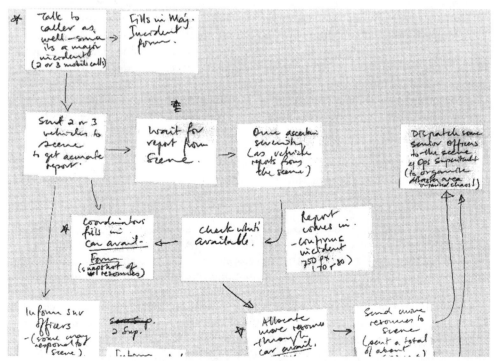

Figure 6.   Use of Post-It™ stickers as visualization aids in documenting the decision process.

information was written on Post-It^{TM} papers and randomly arranged on an A3 size sheet of paper (figure 6).

Participants were then invited to help to sequence the events and decision points by rearranging the Post-It^{TM} papers. As this was done, participants would often recall additional information that would then be added to the diagram. The use of the Post-It^{TM} papers was found to be very useful in encouraging the free flow of detailed information since it was readily apparent to the participants that the exact sequence of events could be revised and reordered subsequently. Participants were observed to remove the Post-It^{TM} papers and reposition them or redraw connecting lines when they were incorrectly recorded by the researcher. The use of this simple low-technology aid appears to enhance the value of the CDM although this remains to be more formally evaluated. The decisions were then probed using the revised CDM protocol shown in the appendix.

5.2. *Results and discussion*

The results are described in detail elsewhere (Wong 1996, Wong *et al.* 1996) and will be briefly summarized here. After transcription, the interviews were coded using the following categories: cues, situation assessments, goals, and courses of action. Cues are stimuli with implications for action. They are obtained from a variety of sources such as the coloured job slips, the vehicle status board, radio communications, etc. The situation assessment describes the elements of the situation, i.e. the 'who, what, where, and when' of the incident. The situation assessment refers to what has happened. What follows on from this is 'what needs to be done', which are coded under potential courses of action. These involve considering the use of helicopters as well as various different types of ambulances and personnel. In any complex dynamic decision-making situation, there may be multiple and sometimes conflicting goals. Goals are identified according to their information value, which derives from the 'context of alternatives' (Olson 1970). For example 'to save lives' would not be identified as a goal since all actions in this context are ultimately directed to this end and therefore the informational value of this statement is low.

The initial goal state was invariably 'to obtain an early and accurate picture of the incident'. The following four goals were repeatedly identified in this study:

I: Maintain situational awareness
II: Match available resources to the needs of the situation
III: Get help on the way within three minutes
IV: Maintain record of events

The goal 'maintain situational awareness' was also of central importance in the study of air traffic controllers carried out for the FAA and described by Redding and Seamster (1994). This led to the development of a training curriculum that emphasized the tasks involved in developing and maintaining situational awareness. In the present case, the aim was to redesign the information display system. The fundamental assumption of the subsequent information design process was that human performance could be improved if information is portrayed in a manner that supports the decision strategies invoked to achieve the goal states of the process being controlled (Wong *et al.* 1997). A brief description of this process with respect to goal II (matching resources to needs) is given below.

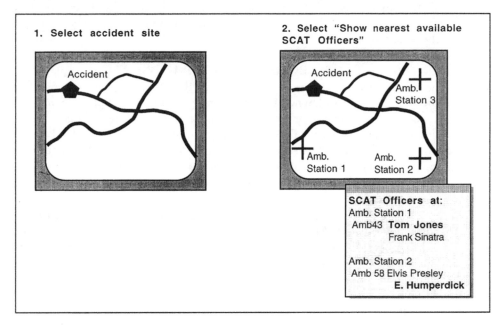

Figure 7. A redesigned ambulance dispatch display to support the operator goal of matching available resources to the needs of the situation.

5.3. *Design implications*

Matching the resources required by the task to those available involves several processes. In the current system, locating specialized personnel was done by reference to hard copy files of daily staffing schedules. Having identified which specialized personnel were on duty, the dispatcher was then required to determine which of these officers was closest to the scene. This inefficient multi-stage planning process could be improved by a single-step display such as that shown in figure 7.

The display makes use of current Graphic Information Systems (GIS) technology to identify the ambulance stations that are nearest in travelling time to the event. The display could show only those stations with the required specialist personnel. At the same time as organizing the dispatch of the required resources to the current event, the dispatcher must be concerned with the state of the resources available for future events. A major incident may require all the available ambulances from a given area leaving that area depleted. Other ambulances need to be re-deployed to 'cover the gaps'. To enable this goal to be fulfilled, the display design should support such planning by portraying the resources available, rather than simply those in current use. If ambulances are in current use, but on cases of low-severity, then the display should indicate this directly so that the dispatcher could draw on these resources in the case of high-severity incidents. As Klein (1993: 107) points out, 'The decision requirements can show what needs to be displayed, but they cannot show how it should be displayed'. There are numerous guidelines in the display ergonomics literature (cf. Wickens 1992), which may be used to determine how to best implement these requirements.

A similar approach has been followed with respect to the other goals identified by the CDM yielding a number of display recommendations. Follow-up work to evaluate the acceptability and utility of these recommendations is now under way.

6. Evaluation of the CDM

The key feature of the CDM form of CTA lies in the use of a series of cognitive probes. There is a crucial difference between simply asking participants 'what did you think about at this decision point?' and the careful probing of cues, goals, expectancies, etc generated by the CDM. In the first case, the response is likely to be a blank look or at best a reconstruction of what would seem in hindsight to have been an appropriate line of thought at the time of the event. The three studies above demonstrate the additional insights into the decision process provided by the CDM.

The present paper thus supports and extends the original work of Klein *et al.* (1989) in a number of ways. Methodologically, the authors have presented a revised set of CDM probes (appendix), which they have found useful in generating information about expert performance in a wide variety of tasks. The probe questions are more detailed than those of Klein *et al.* (1989) and incorporate additional distinctions in the study of expertise that have emerged since the original probes were published. The authors have also suggested the use of a simple graphic technique (see §5) involving Post-It^{TM} stickers for constructing the event time-line. The authors have also shown (see §4) how the information generated by the CDM can be structured for instructional systems development by use of the conceptual graph analysis (CGA) approach.

The authors have extended the applications of the CDM from those outlined by Klein *et al.* (1989) to the development of multi-media tools for the evaluation and training of expert performance, and to the specification of system constraints to support the design of computer-based control systems. All three applications involve settings where decisions must be taken within a short space of time in response to critical environmental cues. Yet, the traditional approach to training decision-making in these domains has been to focus on general analytical decision-making frameworks (see Jensen (1995) for aviation examples), which research (cf. Klein and Hoffman 1993, Wiggins and O'Hare 1995) shows to be more characteristic of novice/intermediate decision-making than of expert performance. This reflects the dominance of analytical decision models in the psychological literature and the relative invisibility of perceptual-recognitional models, such as the RPD model. The importance of a theoretically-driven approach to CTA and subsequent system improvements through training and design is briefly addressed in the final section.

7. General discussion

The three case studies reviewed in this paper are designed to illustrate the oft-quoted dictum of Kurt Lewin that 'there is nothing so practical as a good theory'. Both training and system design require the considered integration of theory and practice. As Cannon-Bowers *et al.* (1991: 289) point out, the connections between the two are 'sometimes overlooked or dismissed as irrelevant . . . research that links theory and technique holds great potential for advancing both the science and the application'.

CTA in general and the CDM in particular are examples of the productive links outlined by Cannon-Bowers *et al.* (1991). Derived from the RPD model of decision-making in naturalistic environments, the CDM yields theoretically meaningful insights into decision-making processes and, as shown in study 1, the application of the CDM can generate substantive theoretical insights—in this case the role of 'pre-priming' in situations requiring extremely rapid decision-making. Klein and Hoffman (1993) have discussed the role of mental simulation in evaluating a potential course of action, and also in general situational assessment. The research described in study 1 involves a setting where almost instantaneous decisions are required. This is similar to the setting investigated by McLennan and Omodei (1996), namely football umpiring. The ability of expert referees and rafting guides to produce generally satisfactory decisions in such circumstances suggests a mechanism, labelled pre-priming by McLennan and Omodei, which makes the most plausible situational assessment prototype instantly available at the moment a decision is required. Further research could usefully be directed to establishing the nature of such a process.

Based on the increasingly well-supported theory of recognition-primed decision-making, the CDM offers a coherent view of decision-centered training and design (Klein 1995). In all three studies it was found that expert decision-makers were behaving in ways described by the RPD model, especially with regard to serial rather than concurrent evaluation of options. These studies lend further support to the use and development of the RPD model for understanding and improving naturalistic decision-making.

The authors have described three examples in which training and system design requirements have been supported by information on expert performance derived from the CDM. The perceptual basis of RPD theory is especially compatible with time-pressured dynamic decision-making tasks that have hitherto been viewed as complex problems in multi-attribute utility assessment. By providing a common theoretical basis for defining both system design and training requirements, the CDM reinforces the inherent complementarity of the two processes towards enhancing system effectiveness.

References

ANDERSON, J. R. 1987, Skill acquisition: compilation of weak-method problem solutions, *Psychological Review*, **94**, 192–210.

BEYNON-DAVIES, P. 1995, Information systems 'failure': the case of the London ambulance service's computer aided dispatch project, *European Journal of Information Systems*, **4**, 171–184.

BRANSFORD, J. B., FRANKS, J. J., VYE, N. J. and SHERWOOD, R. D. 1989, New approaches to instruction because wisdom can't be told, in S. Vosniadou and A. Ortony (eds), *Similarity and Analogical Reasoning* (Cambridge: Cambridge University Press), 470–497.

BROCK, J. F. 1997, Computer-based instruction, in G. Salvendy (ed.), *Handbook of Human Factors and Ergonomics*, 2nd edn (New York: Wiley), 578–593.

CANNON-BOWERS, J. A., TANNENBAUM, S. I., SALAS, E. and CONVERSE, S. A. 1991, Toward an integration of training theory and technique, *Human Factors*, **33**, 281–292.

CRANDALL, B. and GETCHELL-REITER, K. 1993, Critical decision method: a technique for eliciting concrete assessment indicators from the intuition of NICU nurses, *Advances in Nursing Science*, **16**, 42–51.

DRISKELL, J. E., COPPER, C. and MORAN, A. 1994, Does mental practice enhance performance? *Journal of Applied Psychology*, **79**, 481–492.

DRURY, C. G., PARAMORE, B., VAN COTT, H. P., GREY, S. M. and CORLETT, E. N. 1987, Task analysis, in G. Salvendy (ed.), *Handbook of Human Factors* (New York: Wiley)

ERICSSON, K. A. and SMITH, J. 1991, *Toward a General Theory of Expertise* (Cambridge: Cambridge University Press).

FISK, A. D. and GALLINI, J. K. 1989, Training consistent components of tasks: developing an instructional system based on automatic/controlled processing principles, *Human Factors*, **31**, 453–463.

FLANAGAN, J. C. 1954, The critical incident technique, *Psychological Bulletin*, **51**, 327–358.

GIBSON, J. J. 1979, *The Ecological Approach to Visual Perception* (Boston: Houghton Mifflin).

GORDON, S. E. 1995, Cognitive task analysis using complementary elicitation methods, *Proceedings of the Human Factors and Ergonomics Society 39th Annual Meeting* (Santa Monica, CA: Human Factors and Ergonomics Society), 525–529.

GORDON, S. E., SCHMIERER, K. A. and GILL, R. T. 1993, Conceptual graph analysis: knowledge acquisition for instructional system design, *Human Factors*, **35**, 459–481.

GRAESSER, A. C. and CLARK, L. C. 1985, *Structures and Processes of Implicit Knowledge* (Norwood, NJ: Ablex).

HARWOOD, K., ROSKE-HOFSTRAND, R. and MURPHY, E. 1991, Exploring conceptual structures in air traffic control (ATC), *Proceedings of the Sixth International Symposium on Aviation Psychology* (Columbus, OH: The Ohio State University), 466–473.

JENSEN, R. S. 1995, *Pilot Judgment and Crew Resource Management* (Aldershot: Avebury Aviation).

KERSTHOLT, J. H. 1994, The effect of time pressure on decision making behavior in a dynamic task environment, *Acta Psychologica*, **86**, 89–104.

KLEIN, G. A. 1989, Recognition-primed decisions, in W.B. Rouse (ed.), *Advances in Man-Machine Systems Research*, vol. 5 (Greenwich, CT: JAI Press), 47–92.

KLEIN, G. A. 1993, *Naturalistic Decision Making: Implications for Design* (Wright-Patterson AFB, OH: CSERIAC).

KLEIN, G. A. 1995, The value added by cognitive task analysis, Proceedings of the *Human Factors and Ergonomics Society 39th Annual Meeting* (Santa Monica, CA: Human Factors and Ergonomics Society) 530–533.

KLEIN, G. A. and CALDERWOOD, R. 1988, How do people use analogues to make decisions? *Proceedings of a Workshop on Case-Based Reasoning* (San Mateo, CA: Morgan Kaufmann), 209–223.

KLEIN, G. A. and HOFFMAN, R. R. 1993, Seeing the invisible: perceptual-cognitive aspects of expertise, in M. Rabinowitz (ed.), *Cognitive Science Foundations of Instruction* (Hillsdale, NJ: Lawrence Erlbaum), 203–226.

KLEIN, G. A., CALDERWOOD, R. and MACGREGOR, D. 1989, Critical decision method for eliciting knowledge, *IEEE Transactions on Systems, Man, and Cybernetics*, **19**, 462–472.

MCLENNAN, J. and OMODEI, M. M. 1996, The role of pre-priming in recognition-primed decision making, *Perceptual and Motor Skills*, **76**, 1059–1069.

O'HARE, D. 1990, Pilots' perceptions of risks and hazards in general aviation, *Aviation, Space, and Environmental Medicine*, **61**, 599–603.

O'HARE, D. 1997, Cognitive ability determinants of elite pilot performance, *Human Factors*, **39**, 540–552.

OLSON, D. R. 1970, Language and thought: aspects of a cognitive theory of semantics, *Psychological Review*, **77**, 257–273.

ORASANU, J. and CONNOLLY, T. 1993, The reinvention of decision making, in G. A. Klein, J. Orasanu, R. Calderwood and C. E. Zsambok (eds), *Decision Making in Action: Models and Methods* (Norwood, NJ: Ablex), 3–20.

RASMUSSEN, J. 1986, *Information Processing and Human-Machine Interaction: An Approach to Cognitive Engineering* (New York: North-Holland).

REDDING, R. E. 1992, A standard procedure for conducting cognitive task analysis, report to the Federal Aviation Administration Human Technology Inc, McLean, VA.

REDDING, R. E. and SEAMSTER, T. L. 1994, Cognitive task analysis in air traffic controller and aviation crew training, in N. Johnston, N. McDonald and R. Fuller (eds), *Aviation Psychology in Practice* (Aldershot: Avebury), 190–222.

ROSCOE, S. N. 1980, *Aviation Psychology* (Ames, IA: The Iowa State University Press).

ROSCOE, S. N. 1997, Predicting and enhancing flightdeck performance, in R. Telfer and P. Moore (eds), *Aviation Training: Pilot, Instructor, and Organization* (Aldershot: Ashgate), 195–208.

ROTH, E. M. and MUMAW, R. J. 1995, Using cognitive task analysis to define human interface requirements for first-of-a-kind systems, in *Proceedings of the Human Factors and Ergonomics Society 39th Annual Meeting* (Santa Monica, CA: Human Factors and Ergonomics Society), 520–524.

SCHLAGER, M. S., MEANS, B. and ROTH, C. 1990, Cognitive task analysis for the real (-time) world, *Proceedings of the Human Factors Society 27th Annual Meeting* (Santa Monica, CA: Human Factors Society), 1309–1313.

SEAMSTER, T. L., REDDING, R. E., CANNON, J. R., RYDER, J. M. and PURCELL, J. A. 1993, Cognitive task analysis of expertise in air traffic control, *The International Journal of Aviation Psychology*, **3**, 257–283.

THORDSEN, M., MILITELLO, L. and KLEIN, G. 1993, Determining the decision requirements of complex flight crew tasks, in *Proceedings of the Seventh International Symposium on Aviation Psychology* (Columbus, OH: The Ohio State University), 233–237.

VICENTE, K. J. 1995, Task analysis, cognitive task analysis, cognitive work analysis: what's the difference? in *Proceedings of the Human Factors and Ergonomics Society 39th Annual Meeting* (Santa Monica, CA: Human Factors and Ergonomics Society), 534–537.

WICKENS, C. D. 1992, *Engineering Psychology and Human Performance*, 2nd edn (New York: Harper Collins).

WIGGINS, M. and O'HARE, D. 1995, Expertise in aeronautical weather-related decision making: a cross-sectional analysis of general aviation pilots, *Journal of Experimental Psychology: Applied*, **1**, 305–320.

WILLIAMS, A. 1997, Rapid decision making. Unpublished MSc thesis, University of Otago, Dunedin, New Zealand.

WONG, W. 1996, A critical decision method investigation of the dispatch management process. Technical report, Department of Information Science, University of Otago, Dunedin, New Zealand.

WONG, W., SALLIS, P. J. and O'HARE, D. 1996, A goal-oriented approach for designing decision support displays in dynamic environments, in H. Hasan and C. Nicastri (eds), *Proceedings of OZCHI96, The Sixth Australian Computer Human Interaction Conference* (Downer, ACT: Ergonomics Society of Australia), 43–48.

WONG, W., SALLIS, P. J. and O'HARE, D. 1997, Eliciting information portrayal requirements: Experiences with the Critical Decision Method, in H. Thimbleby, B. O'Conaill, and P. Thomas (eds), *People and Computers XII, Proceedings of HCI '97 Conference of the British Computer Society Special Interest Group on Human–Computer Interaction*, 12–15 August 1997, University of West England, Bristol (London: Springer), 397–415.

Appendix. The revised CDM probes

Goal specification	What were your specific goals at the various decision points?
Cue identification	What features were you looking at when you formulated your decision?
	How did you know that you needed to make the decision?
	How did you know when to make the decision?
Expectancy	Were you expecting to make this type of decision during the course of the event?
	Describe how this affected your decision-making process.
Conceptual model	Are there any situations in which your decision would have turned out differently?
	Describe the nature of these situations and the characteristics that would have changed the outcome of your decision.
Influence of uncertainty	At any stage, were you uncertain about either the reliability or the relevance of the information that you had available?
	At any stage, were you uncertain about the appropriateness of the decision?

Information integration	What was the most important piece of information that you used to formulate the decision?
Situation awareness	What information did you have available to you at the time of the decision?
Situation assessment	Did you use all the information available to you when formulating the decision?
	Was there any additional information that you might have used to assist in the formulation of the decision?
Options	Were there any other alternatives available to you other than the decision that you made?
	Why were these alternatives considered inappropriate?
Decision blocking — stress	Was there any stage during the decision-making process in which you found it difficult to process and integrate the information available?
	Describe precisely the nature of this situation.
Basis of choice	Do you think that you could develop a rule, based on your experience, which could assist another person to make the same decision successfully?
	Do you think that anyone else would be able to use this rule successfully? Why/Why not?
Analogy/generalization	Were you at any time, reminded of previous experiences in which a *similar* decision was made?
	Were you at any time, reminded of previous experiences in which a *different* decision was made?

CHAPTER 11

Task analysis for the investigation of human error in safety-critical software design: a convergent methods approach

N. M. Shryane*, S. J. Westerman, C. M. Crawshaw, G. R. J. Hockey and J. Sauer

Human Factors Group, Department of Psychology, University of Hull, Hull
HU6 7RX, UK

Keywords: Safety critical systems; Task analysis; Error analysis; Human error.

An investigation was conducted into sources of error within a safety-critical software design task. A number of convergent methods of task- and error-analysis were systematically applied: hierarchical task analysis (HTA), error log audit, error observation, work sample and laboratory experiment. HTA, which provided the framework for the deployment of subsequent methods, revealed possible weaknesses in the areas of task automation and job organization. Application of other methods within this more circumscribed context focused on the impact of task and job design issues. The use of a convergent methods approach draws attention to the benefits and shortcomings of individual analysis methods, and illustrates the advantages of combining techniques to analyse complex problems. The features that these techniques should possess are highlighted.

1. Introduction

1.1. *Task analysis for the investigation of human error*

The term 'task analysis' describes a plethora of techniques intended to describe and examine the tasks carried out by human beings within a system (for a comprehensive review see Kirwan and Ainsworth 1992). The range of human factors domains to which task analysis techniques can be applied is broad, including training, task and job design, allocation of function and performance assurance. Although they have the same general goal, different techniques may be suitable for answering different kinds of questions, in different kinds of work systems. This paper is concerned with an investigation into the human factors underlying the commission and detection of human error in railway signalling software design.

Within the context of performance assurance (i.e. the consideration of factors necessary to ensure system performance within acceptable tolerances), human error is of paramount importance, especially in safety-critical systems. Many analysis methods can be used to investigate the role of human fallibility in systems (e.g. SHERPA, human HAZOP; see Kirwan 1992a). Used as Human Error Identification (HEI) techniques, they are often applied within the framework of Probabilistic Risk Assessment (PRA), where the set of undesirable events that could occur within a system is defined, along with the paths that lead to them and the probability of their occurrence. The assessment and use of Human Error Probabilities (HEPs) has been criticized when applying absolute error probabilities (Hollnagel 1993). The authors

---

* Author for correspondence.

would argue that HEPs are more reliable when comparing *relative* error probabilities associated with different parts of a task than when used to give *absolute* error probabilities.

1.2. *Human reliability analysis and the nature of the system*

For process control system operation and similar environments, where many of these techniques were developed, HEI methods have undoubtedly proved to be useful. Such systems are 'physical' in nature and the hazards they present tend to be delimited, or 'bound', by the physical properties of system elements and the actual or potential physical linkages between system components. In this context an operator action—say, closing a valve—will only be able to affect aspects of the system that the valve has the physical potential to affect. Even if this potential includes causing an explosion, the sphere of influence of this is in principle determinable.

These constraints do not apply when considering the design of software systems. Errors made in the task of programming computer-based systems are not subject to the 'bounding' of error by physical principles in the way described above. Instead, the sphere of influence of an error is only limited by the power and flexibility of the programming language used. The development of complex systems requires the use of powerful languages that can express that complexity. This means that even the most simple of errors (e.g. misnaming one variable) could result in unpredictable and potentially disastrous consequences for the logic embodied by the system. These consequences will not necessarily be confined to the specific aspect of the system to which the error relates. By unintentionally overwriting a section of memory, for instance, other—logically and functionally unrelated—parts of the system can be affected. This means that virtually all aspects of the programming task could lead to hazardous outcomes, rather than particular, easily identifiable sections. As stated by Broomfield and Chung (1995: 223) 'no established technique is available for relating software faults to system hazards'.

1.3. *Human error identification methods*

The effectiveness of most HEI techniques depends upon the expertise and experience of the analyst, and this holds true for more general examples of task analysis methods as well. This is because, almost without exception, these methods are based solely or primarily upon the judgements of expert practitioners, who are themselves open to biases and errors of cognition. Whether through interviews or through task documentation, the information gathered regarding tasks and possible errors will usually be subjective in nature. Some HEI methods attempt to reduce the effects of practitioner bias by using expert-system-like computerized question and answer routines (e.g. SHERPA), but the potential problem of a task expert's faulty or incomplete mental model of the system is not addressed. This is of particular importance in the context of computer programming, where there is unlikely to be a complete mental model of how errors will affect system performance for the reasons outlined in § 1.2.

The development of a programming language, e.g. C + +; Ada, is a good example of a complex programming task. The development is usually accompanied or followed by a standardization process, which attempts to remove inconsistencies, undefined behaviour, etc. from the language. The standardization is carried out by committees such as ANSI (American National Standards Institute) and ISO (International Standards Organisation), made up of experts from around the globe.

However, Hatton (1995: 56) points out '[standardisation] committees simply forget things or don't understand the full implications of some features at the time of standardisation'. These errors come to light during use of the languages, and are referred back to the committee by the programmers who discover them. For the Ada language, developed in part for safety-related applications, Hatton (1995) reported that there were around 2000 of these 'interpretation requests' outstanding.

1.4. *Convergent methods approach*

It can be seen that features of the computer software design task pose problems for the investigation of human error. In the main, HEI techniques are powerful and flexible methods that can be used effectively in a range of task environments. It is contended, however, that in the study of error in computer programming, and certainly in the context of safety-critical systems, they should not be used alone. 'It is recommended that reliance is not placed on a single technique of HEI' (Kirwan 1992b: 378). A broader-based approach is needed, using more than one method in order to provide convergent validation, and to allow different parts of system performance to be adequately investigated. This calls for the use of a set of differing analysis methods and data sources, including actual task performance, rather than traditional 'expert-opinion' focused task analysis alone. Using this combination a 'matrix' of evidence regarding overall system integrity can be built up.

This paper considers such an approach to the analysis of human error. Specifically, the design of software for a safety-critical railway signalling control system, called 'Solid State Interlocking' (SSI), is described in § 2. Section 3 details the initial task analysis of this design process. Section 4 describes an empirical error analysis used to provide convergent evidence for the investigation. It should be noted that this study is an investigation into the factors affecting production and detection of error, not a PRA of the system.

2. Solid State Interlocking

As a case study of a complex safety-critical system, the design of data for a railway signalling safety system was investigated. 'Solid State Interlocking' (SSI) automatically controls the signals, points, etc. in a section of railway, ensuring that only safe movements of trains are allowed to take place. Each SSI is triply hardware redundant (three identical central processors), but the 'geographic data' that specifies the logic for the movement of trains is unique to each SSI, and the same version is loaded into all three processors. This means that the 'data' must be correct, as any faults could allow collisions or de-railments.

Figure 1 shows a small section of a highly simplified signalling diagram. It represents a plan view of the railway layout. Track sections are shown labelled T1, T2, etc. Signals are represented by three schematic lights and are labelled S1, S2, etc. Where tracks converge or diverge there are gaps, representing sets of points, labelled P1 and P2.

Below the diagram is a simplified section of data, showing the conditions that must be fulfilled before Route 2 (R2; from S1 to S7) can be set. This entails checking that the route is available (R2 a), e.g. not barred because of maintenance; that the points are in the correct position, or free to be moved to the correct position (P1 crf, P2 cnf); and that other, opposing routes are not already set, which is done by checking two opposing sub-routes to ensure that they are free (U10-AB f, U3-BC f). If these checks are passed, then the route is set (R2 s); the individual sub-routes in

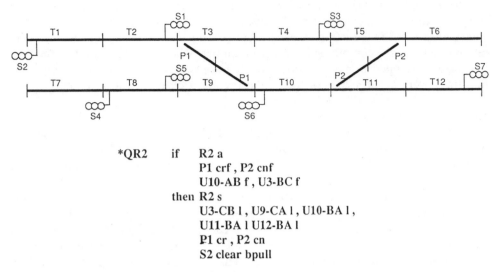

*QR2   if   R2 a
              P1 crf , P2 cnf
              U10-AB f , U3-BC f
        then R2 s
              U3-CB 1 , U9-CA 1 , U10-BA 1 ,
              U11-BA 1 U12-BA 1
              P1 cr , P2 cn
              S2 clear bpull

Figure 1.   Example SSI signalling layout and data.

Route 2 are locked (e.g. U3-CB l); the points are moved to the correct position (P1 cr, P2 cn); and the route entrance signal is checked to see if it is available to be changed to green (S2 clear bpull). It is the programming, verification and validation process for this data that is the focus of the present study.

3.   Task analysis

The technique of Hierarchical Task Analysis (HTA: Annett and Duncan 1967) was chosen for the initial investigation of the SSI design process. HTA has existed for so long, and it has been so widely used, that it could be described as 'traditional'. Associated with it is a range of 'traditional' sources of data, including the technical— and the critical incident—interview, system documentation and operator observation. This technique was chosen because of its flexible and systematic task decomposition process. Used with the data sources listed above, it would provide the framework for the later error analysis, which could then be used to provide a retrospective validation of the technique.

3.1. *Hierarchical task analysis method*

A 'process analysis', as suggested by Piso (1981), was conducted concurrently with the initial stages of the HTA. The process analysis sets out to describe the production process, functioning of equipment, jargon, and general task environment, so that a framework for the HTA itself is provided. HTA is used to describe the goals that the operator must fulfil and the plans to schedule them. First, the overall, or superordinate, goal of the SSI design system is broken down into a number of sub-goals, along with the plan required for structuring, ordering and performing them. This procedure is then iterated, each sub-goal being assessed to see if it warrants further redescription. Consideration of performance shaping factors (e.g. expertise, task design, performance demands, office environment) informed the assessment of error-proneness of the operation or plan, and hence the level of redescription required.

The strengths and limitations of each of the sources of data listed earlier, i.e. interviews with task experts, task-related documentation and video recorded task observation, are discussed below.

3.2. *Sources of data for HTA*

3.2.1. *Interviews*:   Interviews with task experts are an accessible and flexible method of data collection. Interviewing is, however, a purely subjective method, liable to error, omission and bias. To reduce these effects a number of steps were taken. Various personnel were interviewed, from managers to trainees, to avoid gaining a view of the data design task coloured from one organizational perspective. As well as formal interviews, informal chats with staff were entered into wherever possible. It was felt that, especially in a safety-critical industry, the less the participants were made to feel scrutinized the more likely they were to be honest about the way in which tasks were conducted and about the errors that could occur.

3.2.2. *Documentation*:   Guides, manuals and training materials give detailed information on how a task *should* be done, although they are unlikely to reveal how a task is *actually* done. In the domain of programming tasks, it will tend to be more a description of the 'tools of the trade', rather than a 'recipe' to achieve correct task performance.

3.2.3. *Observation*:   Task observation allows features and problems of task performance to be revealed that otherwise may not come to light using the above methods (e.g. assumed knowledge). The 'observer effect' is, however, likely to have an influence on task performance that is especially unwelcome when investigating error.

3.3. *Task analysis results*

The data gathered from all sources were amalgamated into a hierarchical diagram of the SSI data design process. For a fuller description of the HTA, see Westerman *et al.* (1994). The full diagram consisted of seven levels of the hierarchy, around 150 individual operations structured by 40 plans of various complexity. A brief overview of the process of SSI data design is shown in the hierarchical diagram (figure 2), showing the first three levels of the hierarchy.

Box 0 contains the overall goal of the system, boxes 1 and 2 below being the sub-goals required to accomplish it. The relationship between boxes 1 and 2 is shown in Plan 0. These sub-goals are then further redescribed in boxes 1.1 to 2.2. Each horizontal level of the hierarchy is a complete description of the task, at higher detail the lower the level.

The task is roughly divided into two areas (boxes 1 and 2 on the diagram): office-based production and site-based installation. The actual programming process consists of four stages: preparation (writing; 1.2), set to work (simple testing; 1.3), bench checking (code inspection; 1.4) and testing (1.5). These stages will be the focus of this study. In terms of the system life-cycle, these stages correspond to the detailed-design and build of the system (1.2, 1.3), and the verification and validation process (1.4, 1.5). Plan 1 specifies how these tasks are linked. The equipment available for the task is a dedicated UNIX system called the Design Work Station (DWS). The data are written at the terminals of the DWS. Printouts can then be produced for static code inspection, or data checking as it is called. The data can also

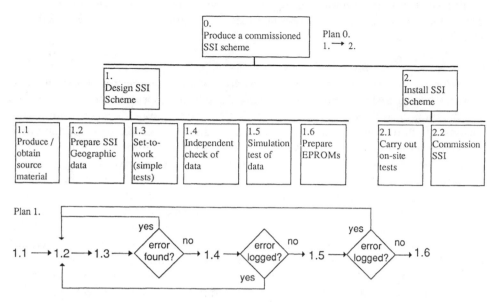

Figure 2.   Overview of the HTA for designing and installing an SSI.

be downloaded to an SSI processor module, which acts as a simulator for the functional testing of the data. The tester performs functional tests of the SSI data by using a trackball to scroll across a graphical representation of the signalling layout (similar to figure 1), and setting elements to their various states.

Essentially, the SSI data are written, and then briefly tested using a computer-based simulation of the railway network. This first testing session is carried out by the author of the data, to ensure that it meets certain minimum standards. A printout of the data is then checked by another engineer who has had no part in the writing process. Finally, the data are again loaded onto a simulator, where they are subjected to more rigorous and formal testing by a specially qualified engineer who must have had no part in either of the preceding stages. If any errors are found during either the check or formal test, they are logged and the data returned to their author for amendments, and the whole cycle gone through once more.

Regarding the staffing for the various stages, expertise tends to increase from left to right. The most expert and qualified signalling engineers are employed at the testing stage, the 'last line of defence' to remove faults in the data. The least experienced are employed in writing the data, although they can seek help from more senior engineers who are not involved on the project. Expertise is gained in the first instance by a number of 2-week training courses that instruct the newly recruited trainee-engineer in basic railway signalling and SSI data. Most training is done 'on-the-job', however, with novices tackling progressively more complicated aspects of the data preparation task, and then going on to qualify for checking, etc. There is not the space here to review all of the potential problems highlighted by the HTA, but some of the more interesting ones are described below.

If the HTA is carried out systematically, structural features of the HTA hierarchical diagram can be used to highlight elements of the task being studied. For instance, the pattern of the overall diagram (not shown) and the specific goals within this pattern revealed the similarity of the writing and checking tasks compared to the testing task.

Consideration of an organizational change revealed a potentially serious problem relating to the introduction of automation in the data writing task. To increase productivity, all of the signalling firms taking part in this study were in the process of developing computer-based tools to automatically write much of the simpler, rule-based data straight from a signalling plan. This will have the effect of removing much of the training that novice data writers gain by tackling these tasks, and leave them less equipped to handle the more complex, knowledge-based data that was identified in interviews as the most problematic.

Time pressure occasionally forces some checking and testing to be carried out in parallel, leading to a revised Plan 1 (figure 3). This means that the version control for the data must be very tight, or unchecked data could be signed-off as safe by the tester. Normally, each new version of the data is given a unique version number by the data writer. This number records how many cycles of checking and testing the data has gone through, but not whether the latest version was generated because errors were found in a check or a test. If it was a test, has that version of data been checked as being error-free before? The danger point is shown by the dashed diamond in figure 3. If this decision is made incorrectly, unchecked data could be released into service. This problem is exacerbated by the contracting-out of the checking or testing of these 'rush' jobs to other signalling firms, with an attendant increase in the difficulty of version control.

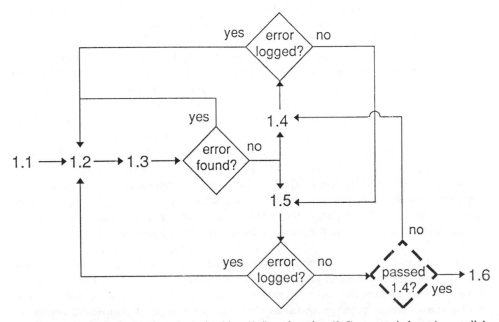

Figure 3.   Revised plan 1, for when checking (1.4) and testing (1.5) are carried out in parallel.

3.4. *The need for error analysis*

The first phases of the analysis showed that HTA provides a useful framework for the breakdown of potential problem areas in a task. As discussed above, structural elements of the hierarchical diagram can be used to show similarities and differences between tasks. However, these similarities do not necessarily equate to similarities in actual task performance. For example, although identified as similar by the HTA, performance in writing and checking may not be identical even given identical data, i.e. data that are difficult to write may be easy to check and vice versa. What HTA does not reveal is how all of the variables that *may* affect task performance will *actually* combine to produce error.

4. Error analysis

The HTA was used to identify the key stages of SSI data production: writing, checking and testing. These were then analysed using a variety of empirical techniques. Crucially, not only the individual tasks in isolation, but also the combination of tasks that together make up the overall system needed to be assessed.

Several complementary techniques were chosen to provide data for the error analysis. These were chosen partly on the basis of availability, but also to give a broad range in terms of the type of data that they would provide. Existing error logs would be supplemented by observation to provide work-based data. A work sample and laboratory experimentation would be used to investigate in detail issues brought to light from the workplace. These methods are now addressed in sequence. (For further details see Westerman *et al.* (1995a).)

4.1. *Error log audit*

4.1.1. *Method*:   The logs used to communicate faults found in formal checking and testing to the data writer were the data source for this method. They revealed errors made not only by the data writer (boxes 1.2 and 1.3 in figure 2), but also by the data checker (box 1.4 in figure 2), as faults found by the tester necessarily have been missed by the checker. These logs detailed the SSI data faults in terms of their functional manifestation rather than their underlying psychological cause, but still provided rich information about the nature of errors. The main strength of this method was its intrinsic validity: the errors were all committed and detected by signalling engineers carrying out their normal job of work.

The primary weakness of the method is the uncontrolled nature of the variables underlying the production and detection of errors. The number of faults in the data depends mainly upon the skill of the data writer and the complexity of the work. Usable measures of these factors were not available, however. In an attempt to reduce bias, logs were included from 12 different signalling schemes, conducted by seven different signalling firms at nine sites. Even when faults within a single scheme were compared, so controlling for expertise and complexity, the detection of a fault by the checker means that the fault is therefore unavailable for detection by the tester.

Two classes of error were not recorded in the logs. The first class is that of the errors committed by the data writer, but also detected by him or her. These would be corrected when discovered, and not be logged and passed on to the later stages. The second class is that containing faults that are not caught by either checking or testing—perhaps the most important to study. Information regarding any faults discovered after the data were installed on-site, or any faults that remained latent until accidents or incidents, was not available.

4.1.2. *Results*:   Table 1 shows the breakdown of 580 faults detected by checking and testing, 290 from each. The 'T/C ratio' specifies the ratio of faults detected by testing versus checking within each category (i.e. testing divided by checking). The higher the figure, the greater is the frequency of detection by testing compared to checking. The nature of the information in the logs means that the fault categories mostly relate to the railway signalling principle that they would violate. The exceptions are the 'None' category, which relates to false-alarm errors by the checker or tester (the data were actually correct); and the 'Other' category, which details miscellaneous SSI data-language specific errors, which do not relate to specific signalling principles.

There are a number of reasons why it would be misleading to use these results to make a formal quantitative comparison of the relative efficacy of the checking and testing processes. First, the error logs were gathered from a number of different schemes and therefore the checking and testing logs were not fully matched. Second, any faults detected at the checking phase were not available for detection at the testing stage, and consequently there is no means of estimating the efficiency of the testing process in detecting these faults. Third, there were no measures available of the total numbers of faults that escaped both checking and testing, again making estimation of the efficacy of the testing stage problematic.

Given these reservations, a number of qualitative features are of note, however. The T/C ratio shows an advantage of checking over testing for the 'Other' category. The 'Opposing locking' category, which contains errors made in barring the setting of two conflicting routes, shows a bias towards testing. Also of note is the preponderance of faults in the 'Identity', 'Other' and 'Route' categories (62.8 % of all faults). Much of the SSI data that relates to these categories is similar to that shown in figure 1. Overall, it is simpler and more straightforward than data controlling other functions, and could be considered as requiring more skill- and rule-based performance and less knowledge-based performance (Rasmussen 1983) than data relating to 'Aspect sequencing', 'Opposing locking', etc.

A total of 12.4% of all faults logged turned out to be false alarms, where no actual fault was present in the data. Several 'repeat' faults were found in the logs. These refer to faults found at either the check or test that were still present when the

Table 1. Faults detected at the checking and testing stages of SSI data production.

Fault category: signalling principle	Checking	Testing	T/C ratio	Category (% of total)
None (false alarm)	46	26	0.57	12.4
Identity and labelling errors	26	30	1.15	9.7
Route setting	74	120	1.62	33.4
Signal aspect control	14	32	2.29	7.9
Approach locking	11	24	2.18	6.0
Opposing locking	9	39	4.33	8.3
Aspect sequence	4	12	3.00	2.8
Other	106	7	0.07	19.7
Total	290	290	1.00	100

SSI data were re-checked or re-tested, and testifies to difficulties in writing the data correctly. These faults applied to particularly novel, knowledge-based SSI data. Where information was available to show the number of faults logged at successive checking cycles, seemingly simple faults made in the 'Identity' and 'Other' categories were the only ones to escape detection until the fourth cycle. Similar information was not available for the testing stage.

4.2. *Error observation*

4.2.1. *Method*: Errors committed but subsequently detected by the same writer, checker or tester were not recorded in the error logs. To compensate for this, a period of *in situ* error observation was conducted. In addition to the task areas accessed by the error logs, this method provided some insight into the fallibility of testers (box 1.5, figure 2). Different engineers were video recorded performing several hours of the three main task areas. Interference of the task could affect error-rate, so to disturb the tasks as little as possible the participants were asked to carry on their work as normal, but to comment if they found that they had, or had nearly, made an error. They were then prompted as to the reasons they identified as causing the error, in an attempt to classify it as a skill- or rule-based slip or lapse, or a knowledge-based mistake (Rasmussen 1983; Reason 1990). The complexity of the task being undertaken, and the expertise and experience of the engineer were also used to inform the categorization of errors. While the possibility of 'faking' a lack of errors existed, the video recording of the sessions and possibility of peer review by other engineers effectively minimized this.

4.2.2. *Results*: Table 2 shows the number of errors flagged by participants during the periods of task observation shown, and is divided up by task stage. The writing task is further subdivided by the nature of the specific writing task being carried out. Data writing does not equate to mere typing (editing), but also includes reviewing, checking specifications and requirements, and planning. It can be seen that data editing on its own accounts for most of the errors in the writing task as a whole. The most error-prone signalling principle in the data editing was 'Opposing locking', with five errors. Of these, the most time consuming of all of the errors was observed. Fully 1.5 hours were spent by one engineer attempting to find the relevant help in the paper-based manual available for the task. In the testing stage, most errors related to confusion over elements selected for testing on the VDU display. No knowledge-based errors were observed in any stage. No errors at all were observed in 3 h of checking.

Table 2. Errors observed at the writing, checking and testing stages of SSI data production.

Error category	Writing (all-including editing) (7.5 h)	Writing (editing only) (3 h)	Checking (3 h)	Testing (6.25 h)
Skill- and rule-based errors	31	25	0	5
Knowledge-based errors	0	0	0	0
Errors per hour	4.1	8.3	0	0.8

4.3. *Work sample test*

4.3.1. *Method*: Since the above two techniques are based on naturalistic analyses of work, they are susceptible to a lack of control over the initial SSI data. To compensate for this, a data writing work sample test was devised that would further inform the analysis of box 1.2 of the HTA (figure 2). A previously completed data set (for an SSI currently in service) had sections of data removed, and signalling engineers were recruited to complete the work under controlled, but work-based, conditions. This enabled task factors (e.g. complexity) to be studied. Fifteen data writers took part in the test. Although this is a small number for statistical purposes, it represents approximately one-third of all of the suitable candidates working for organizations participating in the research project. The participants in the work sample test had a wide range of experience, from 6 months to 10 years, and were drawn from three different signalling offices. To ensure the representativeness of the work sample a highly experienced engineer was employed to select the SSI data, which included all of the main aspects of the writing task. The input of the researchers was used to ensure that both straightforward and novel SSI data elements were represented, so that rule-based and knowledge-based performance could be assumed to be utilized by the participants.

4.3.2. *Results*: Table 3 shows the completion times and faults for the different task types used in the test. For the rule-based task performance, faults in the completed data were scored per signalling function violated, as for the error log audit. This avoided the biasing of the results possible when one cognitive error resulted in many faults in the SSI data. For instance, one error made by a number of engineers, involved the unintentional omission of a whole signalling function. This single cognitive error resulted in the omission of several lines of data and tens of data 'words'. Assessing errors by the line or data 'word' would have given arbitrary scores dependent on how many lines or words made up a particular signalling function.

This scoring system would have been inappropriate for the knowledge-based data, as the small amount of code involved related to only one signalling function, which nobody got completely correct. Additionally, each data 'word' in the knowledge-based section contributed a particular aspect to the overall functionality of the feature, independently of the other 'words' (which was not generally the case for the rule-based data). The knowledge-based task element was thus scored by the data word, to give a usable index of 'correctness'.

Table 3. Completion times and data faults in the work sample test.

Aspect of the task	Completion time (s)			Faults	
	Mean	SD	Time per line	Mean	SD
Rule-based performance (124 lines of data; $n = 15$)	9518	2958	76.8	9.5	7.6
Knowledge-based performance (4 lines of data; $n = 13$)	1120	458	280	3.7	1.8
Total performance (128 lines of data; $n = 13$)	10103	3013	78.9	10.9	4.4

The differences between the two scoring systems makes formal statistical comparison between rule- and knowledge-based fault performances misleading. However, were the rule-based faults to be scored by the data word this would inevitably increase rather than decrease the scores, and so this points to a preponderance of rule-based over knowledge-based faults. On the other had, it can be seen that knowledge-based data is much more time-consuming to complete, per line of data, than the rule-based section (t (12) = 10.7, $p < 0.001$), and its difficulty is also demonstrated by the fact that the two least experienced participants could not complete it. (Their data are thus not included, giving $n = 13$ for the knowledge-based elements.)

Regarding rule-based data, the problem of 'common-mode failure' was highlighted. A term usually used in system safety, it refers to situations in which seemingly independent components can fail due to the same underlying cause. Evidence for common-mode failure was found when it was seen that four specific faults were made identically by seven participants or more; indeed, one of these faults was made by 13 participants. These faults made up the largest single rule-based category, encompassing 40 of the 142 errors.There was no pattern relating to the signalling firms or sites of the participants making these errors, or their expertise. Task-related factors, and their interaction with human cognition, are thus implicated. It was found that three of the four common-mode errors related to familiar data that, because of the specific instances used, required infrequently needed parameters. These errors can be characterized as habit intrusions, or 'strong but wrong' errors (Reason 1990).

4.4. *Laboratory experiment*
4.4.1. *Method*:   In real work, the detection of a fault by a checker renders this fault unavailable to a tester. Therefore, to investigate the relative efficacy of the two processes, computer-based, simplified task simulations were developed for completion by novice participants acting as either checkers ($n = 13$) or testers ($n = 27$). Sets of identical, detectable faults were seeded into both task simulations (16 faults per simulation), and performance of the novice checkers and testers compared (corresponding to boxes 1.4 and 1.5 in figure 2). The fault types were chosen from actual logs in the error log audit. 'Route setting' was chosen because it contained frequent faults. 'Signal aspect control' and 'Opposing locking' showed great differences between checking and testing. Two categories of 'Opposing locking' faults were chosen, reflecting the two different methods by which this signalling principle is dealt with in the SSI data. It was thought that these two methods may have differential effects on the relative efficacy of checking versus testing. Some fault categories (e.g. 'Aspect sequence', 'Other') could not be used because of the need to make the errors equally 'visible' to checkers and testers, and the simplification of the simulation compared to the real task.

4.4.2. *Results*:   Table 4 shows the probability of detection of the four types of faults seeded into the checking and testing task simulations. Analysis of variance revealed no main effect of task type (checking versus testing), but there was a significant interaction between task type and fault type (F (3,72) = 10.58, $p < 0.001$). This is attributable to the comparatively poor performance of checkers in detecting 'Opposing locking I' faults, and of testers in detecting 'Signal aspect control' faults.

Table 4. Probability of fault detection (checking versus testing) in SSI task simulation.

Fault type (4 faults per category)	Checking	Testing
Opposing locking I	0.69	1.0
Opposing locking II	0.83	0.89
Signal aspect control	0.94	0.41
Route setting	0.90	1.0

False alarm rate (i.e. the percentage of logs relating to non-existent faults) was 22.5% in the checking task, 27.6% in the testing task, 25.0% overall.

5. Discussion

The utility of different task- and error-analysis techniques, as applied to the identification and analysis of human error in the SSI design process, is apparent when these methods are considered in concert. This 'triangulation' (Denzin 1988) confers powerful cross-validation of the techniques by consideration of how the evidence presented by each combines when applied to the same task area. It may be that the methods do not address the same factor, in which case the methods are independent. If they do address the same factor, they can either converge or diverge.

5.1. *Error type*

An example of divergence between analysis methods is seen when considering the type of error most problematic for the SSI data production system. HTA critical-incident interviews consistently identified complex, knowledge-based data design tasks as the most problematic type. In support of this, no participants in the work sample were able to complete the knowledge-based data without faults, and the time taken to complete this small amount of data also testifies to its difficulty. In the error logs, novel, complex data were repeatedly passed on to the checking stages while still incorrect, again suggesting the difficulty in writing these data. The error observation technique showed that knowledge-based errors also appeared much less likely than rule-based errors to be self-detected during writing.

On the other hand, the error log audit found that the majority of faults were being discovered in simpler, rule-based SSI data. From the error observation it was seen that around eight skill- and rule-based errors per hour were being self-detected while data editing, and the work sample test showed that around four skill- or rule-based errors per hour were not being self-detected. This indicated that a large number of simple errors were being committed in the data writing, with a significant proportion not being self-detected by their engineer. Although a comparison of errors per hour between faults in the rule- and knowledge-based data in the work sample would not be fair (for the reasons outlined in § 4.3.2), it was shown that, as in the error logs, many more rule-based faults occurred over the whole task.

So knowledge-based data seemed noticeably difficult and error prone to write, but in all tasks viewed and the error logs, rule-based errors and faults predominated. The answer probably lies in the relative opportunities to commit both types of error. The work sample task contained only a small amount of complex, knowledge-based data to complete, but it was selected in part because of its similarity to actual work; indeed, it was part of an actual SSI scheme.

A factor not considered in the above argument is fault detection. The error logs do not provide an accurate ratio of knowledge-based to rule-based faults pre- and post- checking and testing, which would allow the relative detectability of each to be known. However, there is some continuous data (i.e. from the same SSI dataset) relating to faults detected at each stage of checking. This showed that the only faults to pass through three cycles without detection were ones in simple, rule-based data. Again, this may not necessarily be because simpler faults are actually any harder to detect than the knowledge-based ones. Indeed, even if they are easier to detect, the number of simple faults escaping detection may be higher because of their higher prior probability in the data.

Another factor in the number of rule-based faults may be the relative lack of emphasis placed on their importance, as shown by the HTA interviews. This would lead to more attention being given to parts of the data seen as challenging or potentially more error prone than the simple and straightforward aspects. However, virtually any faults in the data could have disastrous consequences.

The HTA suggested that knowledge-based data writing was more complex and made more demands on the operator than simpler data, and this was supported by the work sample test. However, HTA has little provision to represent explicitly the event-rates that would have shown that knowledge-based data, although perhaps more error-prone, is written far less often than the simpler data, and so has correspondingly less influence on overall system integrity.

The prevalence of simple errors and their difficulty in detection has unanticipated consequences for the introduction of automation. Although the original reasons for the development of automation was economic, its introduction may have greater than expected benefits if it can eradicate a potentially dangerous source of data faults. However, its implications for training will still need to be addressed.

5.2. *Common mode error*

Common mode error was another problem area uncovered by the study. Checking, or static code inspection, is seen as the most efficient method for revealing error in computer programming code (Bezier 1990). The principle reason for this is that the whole of the code can be inspected, as opposed to the limited amount of functionality that can usually be tested in complex systems. (The error logs and experiment showed both checking and testing to be similarly efficient; but this is probably because a greater proportion of SSI functionality is testable than for normal programmes.) However, the similarity of checking to the writing stage, suggested by the HTA, may reflect underlying similarities in the tasks and mental processing required by both stages. If this is the case, then checking may suffer the same weaknesses as writing, and so the processes may be liable to common-mode failure. Indeed, a fault type seen to be problematic in the data writing error observation ('Opposing locking'), was detected poorly by checkers in the error log audit and in the laboratory experiment. Further evidence for this similarity is also provided by the fact that the kind of simple errors observed most frequently in the observation were those most resistant to detection by checking in the error log audit. This may seem odd when considering simple slips, as they can be generated by processes not affecting the checkers (e.g. typographic errors), and they are also easiest to self-detect. However, the most striking examples of common-mode error were the identical errors committed in the work sample test, with up to 87% of the engineers making exactly the same error. This highlights the vulnerability of even highly trained engineers to exactly the same errors of cognition when in the same task environment. Checkers will be liable to these same errors because of the similar task factors to writing, and self-detection of these 'habit intrusions' will not be good, as they are errors of intention, not action. Common-mode errors were also seen in the laboratory experiment, with task environment (checking or testing) effectively predicting which types of detection errors would be made.

5.3. *Task diversity*

The converse of task similarity is task diversity. Differences between checking and testing, suggested by the HTA, were confirmed by qualitatively different error detection performance between checking and testing found in the error log audit. While encouraging, the error log results may have been due to a number of factors other than the task environment (e.g. the initial number and type of faults in the data). However, the result was confirmed in the laboratory experiment, where the seeded faults were exactly the same for checkers and testers. The results showed that while there was no difference in the overall fault detection performance between the two methods, they did lead to the discovery of different types of fault. Almost inevitably, because of the differences between the real tasks and the simulations, there were some differences of detail between the error-log and laboratory experiment results, e.g. in Signal aspect control faults. A number of factors may have contributed to this, e.g. the difference in participant characteristics, the reduced range of faults in the simulation (16 versus 580 in the error logs). Given these differences, however, the fact that both studies still showed different fault detection characteristics between checking and testing indicates that rather than the two stages being an example solely of redundancy in the system, they are instead an example of task diversity.

The use of task diversity can have positive implications for fault detection tasks (Fagan 1986). Different task representations are likely to engender different mental models in the operators, and lead to different emphases and task performance strategies. This can render people less vulnerable to the threat of common mode error present when tasks are too similar, as shown between SSI data writing and checking. This diversity of knowledge, strategy and mental model of a task is known as 'Cognitive diversity' (Westerman *et al.* 1995b, 1997).

The diversity between checking and testing has implications for the consequences of the check stage being missed out when carrying out checking and testing in parallel (figure 3). This is because some of the faults in the data, such as those that fall into the 'Other' category, are less likely to be detected by testing alone. To help to ameliorate this problem a computer-based logging tool is currently under development by the research team, to assist with fault logging and data version control. It will also record some of the *psychological* error mechanisms that lead to the faults, and help to aid the identification of further error reduction techniques.

5.4. *'Opposing locking' faults*

A consistent difference between checking and testing performance was found for 'Opposing locking' faults. It was apparent that checkers found these faults more difficult to detect than testers, so a number of further experiments were performed to ascertain why this might be the case (Shryane *et al.* 1996). 'Opposing locking' is dealt with in SSI data by the use of sub-route labels, which define a section of track and also the ends of this section that the train will enter and leave by. For example in figure 1, 'U3-CB' specifies the sub-route 'U3' (corresponding to track section 'T3') and that a train will enter/exit in the 'CB' direction. Exactly what 'CB' means in terms of the associated spatial configuration of the signalling diagram is determined by the particular layout of the track, and is different for different track section shapes. It was found that this inconsistent mapping between the spatial information of the diagram and the textual sub-route label was associated with poor performance in a simulated checking task. Testers do not have to assess the sub-route labels directly, only the signalling functionality associated with their action.

5.5. *Convergent methods*

The methods used in this analysis all had a role to play in the investigation of human error. In the first instance, HTA provided an overview of the task not afforded by the other methods. Additionally, HTA does not have to be a study of the task *as is*, but can be used to investigate the implications of variations in the task. Together, these features allowed the discovery of the organizational issues regarding the introduction of automation. It is also useful for the consideration of sequencing and scheduling of tasks, by nature of its plans, leading to the discovery of problems with parallel checking and testing (figure 3). These issues were not brought to light by the other methods. HTA's weakness was found to be in part due to the subjective nature of its data sources. However, the inability to show quantitative aspects of tasks, such as event and error rate, and how this affects the system is the biggest drawback of HTA when considering the study of error.

The error log audit and error observation are both essentially error sampling procedures of the task. Although the information that they produce can be 'noisy' due to its work environment origin, they do provide quantitative data on event and error frequencies. The error logs did not provide direct evidence of certain types of errors, however (e.g. self-detected errors), and so to compensate for this in the current study, error observation was used. The error observation seemed to be useful in recording errors in relation to overt actions, rather than to covert cognitive processes, as shown by the lack of knowledge-based errors while checking. 'There is some evidence ... that while people are good at catching their own errors of action, they are much less good at catching their own errors of thinking, decision making, and perception' (Senders and Moray 1991: 78).

Error observation was useful for pointing out interface and task support factors. The errors in the data editing part of the writing task point to the error proneness of inputting data through a standard keyboard, which may be reduced by the use of visual programming environments and direct manipulation of code. However, this would then make the data writing task environment more similar to the testing task, so reducing overall system diversity. The system-wide effects of such factors need much further investigation. The inadequacy of existing manuals, the primary reference when needing assistance, was also observed. As a further part of this project, a computer-based 'help' application is being developed. This will include the information contained in the existing paper-based manual, but with improved searching and cross-referencing, and the ability to annotate and personalize the manual to support individual working styles. Job support tools such as these may help to reduce the problem of less training, but again research is needed to see to what extent.

The work sample test and laboratory experiment are both types of experiment. These were needed to study the variables identified as important by earlier stages, e.g. task diversity. Although less controlled than experimentation in the laboratory, the work-based version is more valid in terms of environment, task and participants. However, when studying task-related variables for which task knowledge and experience may matter, the use of naïve participants in the laboratory may be beneficial.

6. Conclusions

From the evidence presented above, two dimensions of variation can be identified with respect to techniques used for the investigation of human error in this study. First, the techniques varied in their capacity to represent event—and therefore error—frequencies; HTA lacks the capacity of the other, more empirical, techniques in this respect. Second, the empirical techniques differ in the familiar trade-off between validity and control. Error logging and observation represent highly externally valid techniques. Laboratory experimentation represents the extreme of control and internal validity, with work sample tests offering characteristics between the two ends of the spectrum. Used here to investigate human error in safety critical systems, analysis of human error will in any work-based system benefit from the application of techniques that vary in these properties.

Acknowledgements

The work described here forms part of the project 'Human Factors in the Design of Safety Critical Systems' (Grant No. GR/J17319) from the UK Safety Critical Systems Research Programme supported by the Department of Trade and Industry and the Engineering and Physical Sciences Research Council. It was completed with the collaboration of Interlogic Control Engineering, GEC Alsthom Signalling Ltd, Signalling Control UK Ltd and Westinghouse Signals Ltd.

References

ANNETT, J. and DUNCAN, K. D. 1967, Task analysis and training design, *Journal of Occupational Psychology*, **41**, 211–221.

BEZIER, B. 1990, *Software Testing Techniques* (Amsterdam: van Nostrand Reinhold).

BROOMFIELD, E. J. and CHUNG, P. W. H. 1995, Using incident analysis to derive a methodology for assessing safety in programmable systems, in F. Redmill and A. Anderson (eds), *Achievement of Assurance and Safety. Proceedings of the Safety-Critical Systems Symposium,* Brighton, February, (London: Springer-Verlag).

DENZIN, N. K. 1988, *The Research Act: A Theoretical Introduction to Sociological Methods,* 3rd edn (Englewood Cliffs, NJ: Prentice-Hall).

FAGAN, M. E. 1986, Advances in software inspections, *IEEE Transactions on Software Engineering*, **12**, 744–751.

HATTON, L. 1995, Programming languages and safety related systems, in F. Redmill and A. Anderson (eds), *Achievement of Assurance and Safety. Proceedings of the Safety-Critical Systems Symposium,* Brighton, February, (London: Springer-Verlag).

HOLLNAGEL, E. 1993, *Human Reliability Analysis: Context and Control* (London: Academic Press).

KIRWAN, B. 1992a, Human error identification in human reliability assessment. Part 1: Overview of approaches, *Applied Ergonomics*, **23**, 299–318.

KIRWAN, B. 1992b, Human error identification in HRA. Part 2: Detailed comparison of techniques, *Applied Ergonomics*, **23**, 371–381.

KIRWAN, B. and AINSWORTH, L. K. 1992, *A Guide to Task Analysis* (London: Taylor & Francis).

PISO, E. 1981, Task analysis for process-control tasks: the method of Annett *et al.* applied, *Journal of Occupational Psychology*, **54**, 247–254.

RASMUSSEN, J. 1983, Skills, rules and knowledge; signals, signs and symbols, and other distinctions in human performance models, *IEEE Transactions on Systems, Man and Cybernetics*, **SMC 13**(3), 257–266.

REASON, J. 1990, *Human Error* (Cambridge: Cambridge University Press).

SENDERS, J. W. and MORAY, N. P. 1991, *Human Error: Cause, Prediction, and Reduction* (Hillsdale, NJ: Lawrence Erlbaum).

SHRYANE, N. M., WESTERMAN, S. J., CRAWSHAW, C. M., HOCKEY, G. R. J. and WYATT-MILLINGTON, C. W. 1996, The influence of task difficulty on performance in a safety-critical labelling task, in A.F. Özok, and G. Salvendy (eds), *Advances in Applied Ergonomics* (West Lafayette, IN: USA Publishing).

WESTERMAN, S. J., SHRYANE, N. M., CRAWSHAW, C. M. and HOCKEY, G. R. J. 1995a, Error Analysis of the solid-state interlocking design process: Report no. SCS-04, Department of Psychology, University of Hull.

WESTERMAN, S. J., SHRYANE, N. M., CRAWSHAW, C. M. and HOCKEY, G. R. J. 1997, Engineering cognitive diversity, in T. Anderson and F. Redmill (eds), *Safer Systems. Proceedings of the Fifth Safety-Critical Systems Symposium*, Brighton, February, (London: Springer-Verlag).

WESTERMAN, S. J., SHRYANE, N. M., CRAWSHAW, C. M., HOCKEY, G. R. J. and WYATT-MILLINGTON, C. W. 1995b, Cognitive diversity: a structured approach to trapping human error, in G. Rabe (ed.), *Proceedings of the 14th International Conference on Computer Safety, Reliability and Security, Belgirate, Italy,* October (London: Springer-Verlag), 142–155.

WESTERMAN, S. J., SHRYANE, N. M., SAUER, J., CRAWSHAW, C. M. and HOCKEY, G. R. J. 1994, Task analysis of the solid-state interlocking design process, Report no. SCS-01, Department of Psychology, University of Hull.

CHAPTER 12

Learning to predict human error: issues of acceptability, reliability and validity

NEVILLE A. STANTON* and SARAH V. STEVENAGE

Department of Psychology, University of Southampton, Highfield, Southampton
SO17 1BJ, UK

Keywords: Human error; Methodology; Reliability; Validity; Sensitivity.

Human Error Identification (HEI) techniques have been used to predict human error in high risk environments for the past two decades. Despite the lack of supportive evidence for their efficacy, their popularity remains unabated. The application of these approaches is ever-increasing, to include product assessment. The authors feel that it is necessary to prove that the predictions are both reliable and valid before the approaches can be recommended with any confidence. This paper provides evidence to suggest that human error identification techniques in general, and SHERPA in particular, may be acquired with relative ease and can provide reasonable error predictions.

1. Introduction

1.1. *Introduction to human error*

Human error is an emotive topic and psychologists have been investigating its origins and causes since the dawn of the discipline (Reason 1990). Traditional approaches suggested that errors were attributable to individuals. Indeed, so-called 'Freudian slips' were treated as the unwitting revelation of intention: errors revealed what a person was really thinking but did not wish to disclose. More recently, cognitive psychologists have considered the issues of error classification and explanation (Senders and Moray 1991). The taxonomic approaches of Norman (1988) and Reason (1990) have enabled the development and formal definition of several categories of human error (such as: capture errors, description errors, data driven errors, associated activation errors and loss of activation errors) while the work of Reason (1990) and Wickens (1992) attempts to understand the psychological mechanisms that combine to cause errors (such as failure of memory, poor perception, errors of decision making and problems of motor execution). Reason (1990), in particular, has argued that one needs to consider the activities of the individual if one is to be able to identify what may go wrong. Rather than viewing errors as unpredictable events, this approach regards them to be wholly predictable occurrences based upon an analysis of an individual's activities. Reason's definition proposes that errors are:

> those occasions in which a planned sequence of mental or physical activities fail to achieve its intended outcome, [and] when these failures cannot be attributed to the intervention of some chance agency. (Reason 1990:9)

---

* Author for correspondence.

If errors are no longer to be considered as random occurrences, then it follows that we should be able to identify them and predict their likelihood. The impetus to achieve this has been fuelled in the wake of several recent and significant incidents, most notably in the nuclear industry, and there now exist several human error identification (HEI) techniques. What is lacking up to now is a rigorous quantitative and qualitative evaluation of these techniques. The present paper addresses the issue of prediction of error form and readers interested in the separate issues of error likelihood and error occurrence are referred to Kirwan (1994) for a recent review of the area.

1.2. *Predicting human error*
Most HEI techniques work in a similar manner (Kirwan 1992a). First, the activity needs to be broken down into a sequence of sub-stages by, for example, using hierarchical task analysis (Annett *et al.* 1971, Shepherd 1989, Kirwan and Ainsworth 1992). Second, possible and plausible errors are identified for each of the sub-stages in turn. Third, the HEI technique may indicate the possible psychological error mechanisms underlying these identified errors, and finally, the HEI technique may be able to specify a recovery or error reduction pathway. Despite the number of HEI techniques that have been developed, the area lacks clear and informative documentation, and there is increasing concern amongst professionals about the reliability, validity and appropriateness of each HEI technique for different situations. Validation is clearly an important issue and the purpose of the current paper is to develop a combination of quantitative and qualitative procedures for assessing these techniques.

1.3. *Validation studies*
To date, there are very few studies that have attempted to validate HEI techniques (Whalley and Kirwan 1989, Williams 1989, Kirwan 1992a,b, Kennedy 1994, Baber and Stanton 1996). For instance, Whalley and Kirwan (1989) evaluated six HEI techniques (Heuristics, PHECA, SRK, SHERPA, THERP, HAZOP) for their ability to account for the errors known to have contributed to four genuine incidents within the nuclear industry. More recently, Kirwan (1992b) has developed a comprehensive list of eight criteria to evaluate the acceptability of these techniques at a more qualitative level. In an unpublished study, Kennedy (1994) has included Kirwan's criteria when examining the ability of the techniques to predict ten actual incidents retrospectively. While these studies failed to identify a clear favourite from amongst these HEI techniques, all three studies indicated impressive general performance using the PHECA method (now unavailable). From the remaining techniques, SHERPA achieved the highest overall rankings and Kirwan (1992b) recommends a combination of expert judgement, together with the SHERPA technique as the most valid approach.

The strength of these studies lies in the high level of ecological or face validity that they achieve. The methodologies make use of the opinions of expert assessors for the prediction of errors contributing to real-world events. However, these studies do raise several methodological concerns. Specifically, the number of assessors using each technique is small (typically 1 to 3) and the equivalence of assessment across techniques is brought into question because different people are assessing each HEI technique. A second methodological concern centres on the use of subjective rating scales. It is doubtful that the assessors will share the same standards when rating the acceptability or usefulness of an HEI technique. This factor combined with the small number of assessors for each technique means that these data should be accepted with some degree of caution.

In the light of these criticisms, Baber and Stanton (1996) aimed to provide a more rigorous test of the predictive validity of SHERPA. Predictive validity was tested by comparing the errors identified by an expert analyst with those observed during 300 transactions with a ticket machine on the London Underground. The analyses reveal that SHERPA identified 12 error types, with only 3 of these representing false alarms (errors that were predicted but not observed). Following from Signal Detection Theory (Macmillan and Creelman 1991), these values permit the calculation of an index of sensitivity, and Baber and Stanton (1996) suggest that SHERPA provides an acceptable level of sensitivity based on the data from two expert analysts ($r = 0.8$). The strength of this latter study over Kirwan's is that it reports the use of the method in detail as well as the error predictions made using SHERPA. However, it would be interesting to examine the performance of SHERPA using a larger pool of analysts and using novice analysts to examine the important issue of ease of acquiring the method. These issues are incorporated into the present study. In addition, the current paper seeks to develop a more rigorous statistical procedure for general HEI evaluation, by borrowing from the standards used in the development of psychometric tools (Cronbach 1984, Aitkin 1985). Indeed, recent exercises in comparing the performance of psychometric tests illustrates that the standards are generally quite high (Bartram *et al.* 1992, 1995).

1.4. *The present study*

Reliable and valid evaluation procedures rely critically on the use of unbiased judgements. Kirwan's (1992b) work has done much to underline the importance of evaluation, however, it is conceivable that the judgements underlying his eight criteria may be affected by factors such as user opinion (is the user a good judge of a technique?), user experience (opinions may vary with experience of the technique), user familiarity (judgements may be biased by personal preference of one technique rather than by its usefulness), and the masking effects of face validity (an HEI technique may look applicable at a superficial level but still may not do a good job of predicting human error). These same concerns are evident in the psychometric literature (Blinkhorn and Johnson 1990), and consequently, the present study places more emphasis on the quantitative aspects of learnability, reliability and validity.

1.5. *Learnability*

Shackel (1990) defines learnability as the amount of training time taken to achieve the defined level of effectiveness. Potentially, there are three measures of learnability: training time, performance after successive training sessions, and subjective assessment of the ease of acquisition of the technique. The present study held training time constant by making use of group training sessions (for real-world comparability). Performance of the analysts was assessed after each training session and, in addition, the subjective assessment of ease of acquisition was recorded. On the basis of Kirwan's (1992b) results, it is anticipated that the subjective reports will indicate the relative ease of acquisition of SHERPA.

1.6. *Test-retest reliability*

Test-retest reliability is defined as the correlation of performance on the HEI technique at time 1 with the performance obtained by the same group of people retested after an intervening period of time (Cronbach 1984, Aitkin 1985). Although there are other forms of reliability (Baber and Stanton (1996) discuss alternate forms of reliability in their study) it was felt that test-retest reliability offered a convenient measure of the consistency of HEI techniques used on separate occasions. Based upon the evidence presented by Kirwan (1992b) and Baber and Stanton (1996) the reliability of SHERPA was expected to be quite high. In the use of psychometric tests, a high level of reliability is considered to be essential. The authors believe that this should be the case with HEI techniques also.

1.7. *Concurrent validity*

There are a number of methods for establishing validity of HEI techniques (Cronbach 1984, Aitken 1985). For the purposes of this study concurrent validity was tested. Concurrent validity measures the relationship between the predicted errors using the HEI technique and errors observed on the task. It is rather like predictive validity, but without the time lag between prediction and observation of criterion behaviour. From previous studies the authors expected a reasonable level of concurrent validity (Kirwan 1992b, Baber and Stanton 1996). In the field of personnel selection, a validity coefficient of above 0.4 is considered to be acceptable, a coefficient above 0.6 is considered to be good, and a coefficient above 0.8 is excellent (Cook 1988).

1.8. *Subjective user opinion*

Finally, the present study also made use of a selection of Kirwan's (1992b) criteria of subjective acceptability of the HEI technique (listed below). From Kirwan's study it is expected that SHERPA would score favourably in terms of its acceptability.

(1) *Comprehensiveness*: this criterion is concerned with the breadth of coverage of the technique, i.e. does it identify all errors or only a specific subset?
(2) *Accuracy*: this could also be called concurrent or predictive validity, i.e. the correlation between errors predicted by the technique and errors observed to occur in practice.
(3) *Consistency*: this criterion is the same as inter-rater reliability, i.e. the degree to which two analysts make the same error predictions.
(4) *Theoretical validity*: this refers to the extent to which the method has a founding in error theory or error construct.
(5) *Usefulness*: Kirwan judged this criterion to be independent of the criterion of theoretical validity: an HEI method does not need to be high on theoretical validity to be considered useful.
(6) *Acceptability*: this refers to the overall degree to which people find the process and the outcome acceptable. This seems to be analogous to content and face validity.

In summary, the aims of the study were to investigate in a rigorous manner the confidence that can be legitimately placed in SHERPA. Methodological concerns with previous studies led the authors to adopt the approach normally associated with testing reliability and validity of psychometric techniques. The study also provided the authors with the opportunity of examining the ease with which SHERPA may be acquired by novices. No studies of this nature have been reported in the literature to date.

2. Experiment 1

2.1. *Method for experiment 1*

2.1.1. *Participants*: Two groups of participants were involved in this study. The first group consisted of 62 undergraduate students aged 19–45 years (modal age, 20 years). Of these 44 were female and 18 were male. These participants formed the control group and received no human error identification (HEI) training.

The second group consisted of 36 participants drawn randomly from an equivalent pool as above. These participants acted as novice analysts using a method known as Systematic Human Error Reduction and Prediction Approach (SHERPA; Embrey 1986). All participants were equally familiar with the machine upon which the human error analysis was conducted.

2.1.2. *Materials*: No materials were provided for the control participants. However, all SHERPA analysts were provided with three items. First, they received a hierarchical task analysis (HTA) chart describing the action stages involved when using a vending machine to obtain a bar of chocolate (figure 1). Second, they received a written classification of errors that could be applied to the HTA (figure 2). Finally, participants were provided with a pro forma for recording their error predictions. The

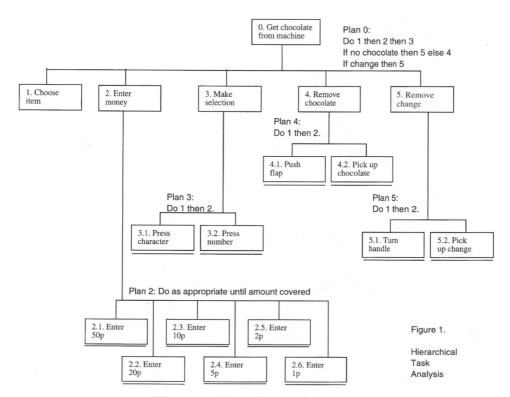

Figure 1.   Hierarchical task analysis.

external provision of the HTA can be justified on two counts: first, the methodological focus of the present research is on the ability of the participants to use the SHERPA technique rather than on their ability to construct an HTA chart; second, as Kirwan and Ainsworth (1992) suggest, in the real world it is quite reasonable for the error analyst to receive an HTA chart constructed by another individual.

2.1.3. *Procedure*:   For both groups, participants were given the scenario of buying one item (a Lion Bar, costing 24p) from the vending machine using a 50p coin and thus requiring change. They were required to try to predict the errors that would occur during this operation and, to this end, all participants received training by means of a 2-h lecture and video on human error. The training began with a general introduction to human error research based upon the work of Reason (1990). A classification system for analysis of human error was presented to distinguish between slips, lapses and mistakes. These error types were defined in terms of an

Error Mode:
Errors are assigned to categories of action, checking, retrieval, communication and selection errors. Credible errors are identified in this column. The coding is as follows:

Action Errors:
A1   Operation too long/short
A2   Operation mistimed
A3   Operation in wrong direction
A4   Operation too little/much
A5   Misalign
A6   Right operation on wrong object
A7   Wrong operation on right object
A8   Operation omitted
A9   Operation incomplete
A10   Wrong operation on wrong object

Checking Errors:
C1   Check omitted
C2   Check incomplete
C3   Right check on wrong object
C4   Wrong check on right object
C5   Check mistimed
C6   Wrong check on wrong object

Retrieval Errors:
R1   Information not obtained
R2   Wrong information obtained
R3   Information retrieval incomplete

Communication Errors:
I1   Information not communicated
I2   Wrong information communicated
I3   Information communication incomplete

Selection Errors:
S1   Selection omitted
S2   Wrong selection made

Figure 2.   Error classification taxonomy.

Information Processing Model (Wickens 1992) and examples of each error type were discussed in various contexts. In particular the link was made between product design and human error (Norman 1988, Thimbleby 1991). This was followed by a 45-min video on human error, which related everyday errors to those errors found in a more unforgiving environment (i.e. the errors contributing to the Tenerife runway disaster of March 1977). Finally, participants using the SHERPA technique received specific instructions in the use of the technique via a 1-h training session. This comprised an introduction to hierarchical task analysis and an explanation of the staged approach of the SHERPA technique as outlined earlier. A worked example was provided and participants then proceeded to generate their own analysis of errors using a familiar everyday device (a kettle).

2.1.4. *Error prediction*: Participants in the control group were required to indicate the errors that they thought would occur during this scenario. No formal training was given and participants had to rely on their own experience and knowledge of the machine.

Participants using the SHERPA method of error prediction received verbal and written training in the use of the method. SHERPA was originally designed to assist people in the process industries (e.g. conventional and nuclear power generation, petrochemical processing, oil and gas extraction, power distribution, etc., Embrey 1986). In essence the SHERPA technique consists of a series of decision stages (figure 3) and works by indicating which error modes are credible for each task step in turn, based upon a hierarchical analysis of work activity. This indication is based upon the judgement of the analyst, and requires input from a domain expert to be realistic.

(1) *Step 1: Hierarchical Task Analysis (HTA)*
    The process begins with the analysis of work activities, using Hierarchical Task Analysis. HTA (Annett *et al.* 1971, Shepherd 1989, Kirwan and Ainsworth 1992) is based upon the notion that task performance can be expressed in terms of a hierarchy of goals (what the person is seeking to achieve), operations (the activities executed to achieve the goals) and plans (the sequence in which the operations are executed). The hierarchical structure of the analysis enables the analyst to progressively redescribe the activity in greater degrees of detail. The analysis begins with an overall goal of the task, which is then broken down into subordinate goals. At this point, plans are introduced to indicate in which sequence the sub-activities are performed. When the analyst is satisfied that this level of analysis is sufficiently comprehensive, the next level may be scrutinized. The analysis proceeds downwards until an appropriate stopping point is reached (see Annett *et al.* (1971) and Shepherd (1989) for a discussion of the stopping rule).
(2) *Step 2: Task classification (completed by participants)*
    Each operation from the bottom level of the analysis is taken in turn and is classified from the error taxonomy, into one of the following types:
    (a) Action (e.g. pressing a button, pulling a switch, opening a door)
    (b) Retrieval (e.g. getting information from a screen or manual)
    (c) Checking (e.g. conducting a procedural check)
    (d) Selection (e.g. choosing one alternative over another)
    (e) Information communication (e.g. talking to another party)

(3) *Step 3: Human Error Identification (HEI) (completed by participants)*
    This classification of the task step then leads the analyst to consider credible
    error modes associated with that activity. For each credible error (i.e. those
    judged by a subject matter expert to be possible) a description of the form
    that the error would take is given.
(4) *Step 4: Consequence analysis*
    Considering the consequence of each error on a system is an essential next
    step as the consequence has implications for the criticality of the error.
(5) *Step 5: Recovery analysis*
    If there is a later task step at which the error could be recovered, it is entered
    next. If there is no recovery step then 'None' is entered.

2.1.5. *Error classification*: The error predictions from all participants were
compared to the errors actually observed in 75 independent transactions with the
machine. Observation of the 75 transactions revealed 11 error types and it was
possible for more than one error type to occur within a single transaction. These
error types are listed in the appendix and errors were identified by a combination of
observation and interview with the users of the machine. The transactions were
observed without the prior knowledge of the user and these 75 transactions provided
a sample of errors that contained all the error types that were likely from a larger set
of observations. In an independent study by Baber and Stanton (1996) it was shown
that a data set of over 300 person-machine interactions revealed 90% of the error
types within the first 20 interactions. Moreover, no novel error types were evident
after 75 interactions. The comparison of predicted and observed errors yielded three
dependent variables:

(a) hits (predicted errors that were seen to occur);
(b) false alarms (predicted errors that did not occur); and
(c) misses (errors that occurred but were not predicted).

The frequency of misses was obtained by subtracting the number of hits from the
total number of errors observed ($n = 11$). These three dependent variables formed
the basis for subsequent analyses.

2.2. *Results for experiment 1*
For each participant, the frequency of hits, misses and false alarms when predicting
errors with a vending machine were calculated. Table 1 summarizes these data across
the control group and the group using the SHERPA method for human error
identification.

From table 1, it appears that the participants using the SHERPA technique
correctly predicted more errors and missed fewer errors. However, they also
appeared to incorrectly predict more errors. Three independent samples *t*-tests
examined whether these differences were significant. The results revealed a significant
difference in hit rate, and also in miss rate, ($t = 18.37$, df $= 95$, $p < 0.001$) in favour of
participants using the SHERPA method. However, the results also suggested that
these participants made significantly more false alarm predictions as well ($t = 19.05$,
df $= 95$, $p < 0.001$). There appears to be a trade-off in terms of training such that a
more sensitive human error identification is achieved, at the cost of a greater number
of false positives. This is probably a conservative estimate, as no doubt if the

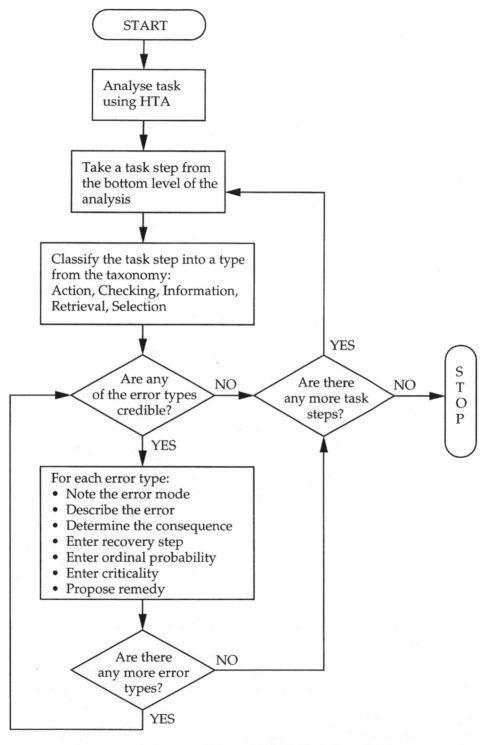

Figure 3. Decision stages involved in the SHERPA technique.

Table 1. The mean frequency of hits, misses and false alarms for SHERPA and untrained participants.

	Untrained participants ($n = 62$)		SHERPA participants ($n = 36$)	
	Mean	SD	Mean	SD
Hits	2.4	1.2	8.2	1.9
Misses	8.6	1.2	2.8	1.9
False alarms	0.5	0.7	15.4	6.1

observation period was extended indefinitely, more error types would be observed eventually.

2.3. Discussion for experiment 1

These results suggest that when participants first use the SHERPA method of human error identification, they are able to correctly predict more errors (and hence, miss fewer errors) than participants who have had no training. In this respect, some training does appear to be better than none.

However, in two respects, the SHERPA method appears worse than no training. First, from the verbatim reports it is clear that some possible errors that can be predicted heuristically are missed by the SHERPA method. These include (1) entering money without observing the 'exact change' illuminated sign and thus being unable to get change, (2) getting one's hand stuck by the flap when retrieving the chocolate, (3) making a selection when the first slot of the desired line is empty, and (4) pressing the digit '1' and then looking for the digit '0' instead of pressing the digit '10'. This latter error can be captured by the category 'press wrong number' but it seems that this particular error represents a design flaw in the machine because users do not expect there to be one button for a 2-digit number. In this respect it is important to be able to identify such errors independently so that design changes can be implemented.

The second drawback highlighted by the present results is that the participants using the SHERPA method make a greater number of erroneous error predictions than the untrained participants. Indeed, those using the SHERPA method predict more errors *per se* compared to the untrained participants. In terms of the usefulness of a human error identification technique this is not always a desirable outcome. Predicting errors that do not actually occur can lead to needless changes in the design of a machine and this is obviously wasteful in terms of time and financial resources. Consequently, there is a trade-off to be considered between predicting a lot of errors, and 'crying wolf'.

In the light of this it is important to consider the overall sensitivity of the technique rather than merely consider the frequency of hits, misses and false alarms independently of one another. The following studies address this issue as their primary aim. As well as focusing on the sensitivity of the SHERPA technique the studies that follow also address the issue of the reliability of SHERPA over time. Baber and Stanton (1996) 'have informal data that suggest a high degree of reliability of SHERPA use between analysts but they have not looked at the issue of reliability of SHERPA use within an analyst (i.e. across time). By using a within-subjects test-retest procedure here it is possible to determine whether there is a learning curve associated with novices' use of the technique or whether the sensitivity measure demonstrates intra-analyst reliability over time. More specifically, these studies allow the examination of the effect of practice on one's performance when predicting errors.

3. Experiments 2 and 3

3.1. *Method for experiments 2 and 3*

3.1.1. *Participants*: Of the 36 participants who used the SHERPA method in the previous study, 31 repeated the error prediction phase of the procedure after a delay of 2 weeks, and 28 repeated the error prediction a third time after a further delay of 5 weeks. Of this latter group only 25 participants could be identified and matched to their previous performance. Consequently, data from three participants were dropped.

3.1.2. *Materials*: As for the SHERPA group in experiment 1 (figures 1, 2 and 3).

3.1.3. *Procedure*:

3.1.3.1. *Error prediction*: The procedure matched that of experiment 1 as closely as possible. However, there were several important differences. First, there were no untrained control participants. Instead, participants acted as controls for themselves in that the sensitivity of their error prediction could be compared within participants over time. A second important methodological difference between the first study and the present ones arises from the fact that across the three occasions, a repeated measures design was used. Consequently, all participants had practice using their error detection method and had immediate feedback in the form of access to the observational data. This was considered to be an important factor given the desire to mirror the procedures used in recent training regimes (Patrick 1992).

Furthermore, while being familiar with the methodological concerns of common method variance, a tradition established in the psychometric arena is to hold as many factors of the testing situation constant as possible. The introduction of different machines would cloud the issue because it would not be possible to determine whether differences in sensitivity of error prediction over time were due to learning, or due to the ease of predicting errors with the various machines. For these reasons, participants made predictions of errors when using *one* particular machine and this allows one to determine the learnability of the error prediction technique without the confusion of error prediction on different machines.

Apart from these conditions of testing, all other methodological details remained unchanged.

3.1.3.2. *Error classification*: As in experiment 1, the frequency of hits, misses and false alarms were computed and compared with predicted error rates. In addition, the frequency of correct rejections (where errors that did not occur were correctly not predicted) was calculated by subtraction of the number of hits, misses and false alarms from a theoretical maximum (7 actions × 10 action errors = 70. For the purposes of this analysis, the actions involved in 'entering money' are counted as one action in line with the scenario of having one 50p piece). The four measures that resulted were entered into the signal detection grid (figure 4).

From these four measures, an index of sensitivity (S) was calculated according to the formula below (from Baber and Stanton 1996). This gives a value between 0 and 1, with higher values indicating greater sensitivity of error prediction.

$$\frac{\left(\dfrac{\text{Hit}}{\text{Hit+Miss}} + \left(1 - \dfrac{\text{False Alarm}}{\text{False Alarm + Correct Rejection}}\right)\right)}{2}$$

The four frequency measures plus this index of sensitivity formed the basis of the subsequent analyses.

3.2. *Results for experiments 2 and 3*

For the purposes of the following section the data from experiments 1, 2 and 3 are pooled in order to examine the effects of time on the validity and reliability of the SHERPA technique. At each of the three times, the frequency of hits, misses, false alarms and correct rejections were recorded and from these an index of sensitivity was calculated. These are summarized in table 2.

From this table it appears that there is very little change over time in the frequency of hits and misses; however, the frequency of false alarms appears to fall over time and consequently the frequency of correct rejections appears to increase. In terms of the overall sensitivity of error prediction, this shows remarkable consistency over time.

Three separate one-way ANOVAs were performed to examine the significance of any effects of time for hits (and therefore misses), false alarms (and therefore correct rejections), and the sensitivity index. From these it appears that time has no significant effect for the frequency of hits (or misses) ($F(2,48) = 2.69$, $p > 0.05$).

Errors Observed

		YES	NO
	YES	hits	false alarm
Errors Predicted			
	NO	misses	correct rejections

Figure 4.   Signal detection grid recording the frequency of hits, misses, false alarms and correct rejections.

Table 2.   The mean frequency of hits, misses, false alarms and correct rejections and the inject of sensitivity of SHERPA over time (using 25 complete cases only).

	Time 1 ($n=25$)		Time 2 ($n=25$)		Time 3 ($n=25$)	
	Mean	SD	Mean	SD	Mean	SD
Hits	8.4	1.9	7.6	1.9	7.5	2.1
Misses	2.6	1.9	3.4	1.9	3.5	2.1
False alarms	14.9	5.8	11.9	5.3	10.1	4.6
Correct rejections	44.2	5.8	47.2	5.4	48.9	4.6
Mean sensitivity	0.76	0.1	0.74	0.1	0.73	0.1

However, regarding the frequency of false alarms and correct rejections a significant main effect of time emerged ($F(2,48) = 6.67$, $p < 0.005$). The table indicates a decreasing linear function for the frequency of false alarms (and a corresponding increasing linear function for the frequency of correct rejections) and these trends are borne out by a series of *post hoc* contrasts (linear trend: $F(1,24) = 11.08$, $p < 0.005$; quadratic trend: $F(1,24) = 0.27$, $p > 0.05$).

Consequently, it appears that relative novice users of the SHERPA technique are able to achieve a consistent level of correct error prediction and, similarly, demonstrate consistency in terms of the number of errors that they actually miss. In terms of the frequency of falsely predicted errors, these are seen to decrease with the more experience one has with the error prediction method and, as a result, the frequency of correct rejections increases.

The final set of analyses examine whether there is any effect of time on the overall sensitivity of the error prediction method. A one-way ANOVA was performed on the sensitivity index and from this no significant main effect of time emerged ($F(2,48) = 0.65$, $p > 0.05$). Examination of the three occasions further reveals a significant correlation between the sensitivity of error prediction at time 1 and time 2 ($r = 0.65$, $n = 25$, $p < 0.001$) suggesting reliability over time. The correlations of sensitivity between time 1 and time 3 ($r = 0.32$, $n = 25$, $p > 0.05$) and between time 2 and time 3 ($r = 0.39$, $n = 25$, $p > 0.05$) also show a degree of agreement but one that does not reach significance. However, taken together, the results reveal considerable statistical reliability of error prediction over time.

3.3. *Discussion for experiments 2 and 3*

The results of experiments 1–3 taken together indicate a significant degree of reliability over time in the use of the SHERPA method of human error identification. Specifically, the ability to correctly predict errors, and the frequency with which errors were missed remained constant over time. This is important as it indicates that the crucial business of error detection can be achieved to a high degree of accuracy, again and again, rather than on a one-off occasion. In a similar vein these results demonstrate that, rather than there being a learning curve, analysts with very little training and experience can perform at a consistently high standard.

This is not to say that experience does not have any effect. Indeed, over the three trials, the incidence of false alarms was seen to fall and the incidence of correct rejections rose correspondingly. On the face of it this indicates improvement with experience, in that the analysts seem to be showing some degree of learning about the most effective and discriminative way of using the SHERPA technique. However, this improvement may also be attributable to a practice effect in that participants may simply be remembering (a) what they wrote the previous time, or (b) the feedback of actual errors observed. In the light of this, it is not clear to what this improvement may be attributed. In this respect, it would have been interesting to see whether the untrained control group used in experiment 1 showed any improvement over subsequent testing sessions. Their improvement could not be attributable to attainment of expertise with any error prediction method but could only be explained in terms of these feedback and practice effects.

In terms of the overall sensitivity of the SHERPA method the results suggest a highly significant degree of consistency across time. In particular, the stability between time 1 and time 2 is notable. However, it must be remembered that the delay between time 1 and time 2 was 2 weeks while the delay between time 2 and time 3 was 5 weeks (and between time 1 and time 3, 7 weeks). Accordingly, this variation in delay may contribute to the lack of a significant correlation between sensitivities at time 1 or 2 and time 3. Nevertheless, it is notable that, again, an impressive degree of sensitivity of error prediction can be achieved and maintained with relatively little training and experience.

All these results indicate that the SHERPA technique for human error identification is simple enough to be used effectively by novices. The high degree of sensitivity indicates that the method demonstrates predictive validity—it achieves what it sets out to achieve. Furthermore, the lack of main effect of time for the frequency of hits and misses and for the overall sensitivity index suggests that the method demonstrates reliability over time—the ability to predict errors remains consistent.

The final aspect of the SHERPA method that needs to be evaluated is its perceived usability. The following experiment addresses this issue by means of a usability questionnaire.

4. Experiment 4

4.1. *Method for experiment 4*

4.1.1. *Participants*:   Twenty-eight undergraduate students who had completed the third testing phase took part in this final evaluation of the SHERPA method.

4.1.2. *Materials*:   All participants were required to fill in an 8-item questionnaire. The questionnaire was devised to assess the SHERPA method on the basis of Kirwan's (1992b) criteria for usability. These include rating the method on perceived accuracy, structure, psychological basis, future use, ease of use, derivation of errors, realism of predictions and finally, ease of learning. For each criterion, participants were asked to indicate their feelings on a 7-point Likert-type scale in which a high score was more favourable.

4.1.3. *Procedure*:   Immediately after completion of the third error prediction phase, participants were required to fill in the usability questionnaire. This was completed in a group setting and took no more than 10 min. All questionnaires were completed anonymously, after which participants were thoroughly debriefed.

4.2. *Results for experiment 4*

Participants were required to rate the SHERPA method of error prediction on a 7-point scale for each of 8 criteria. These ratings are summarized in table 3.

From this table it appears that the participants rated the SHERPA method as particularly easy to use and easy to learn. This is consistent with the quantitative data (see experiments 2 and 3), which suggest that the participants found it quite easy to derive the errors using the method. In addition, they suggest quite a high likelihood that they would use the method again if required to predict errors. However, the ratings suggest that the participants were not wholly convinced of the psychological basis of this method of error prediction. Equally, they were not convinced of the realism of the method in terms of the errors that were predicted. Certainly, if we recall the types of errors that the participants predicted heuristically (experiment 1), then it is clear that the errors that one predicts when thinking common-sensically reflect different concerns. They are more global and concentrate

Table 3. Mean rating of the SHERPA method on eight criteria of usability (where 7 = excellent, and 1 = very poor).

Criterion	Mean rating	SD
How accurate was the prediction of errors?	4.04	1.2
How structured was the method?	4.14	1.5
Is the method psychological in approach?	3.07	1.4
Would you use the method again?	4.07	1.4
How easy was the method?	4.64	1.7
How easily were the errors derived?	4.64	1.3
Were the predictions of errors realistic?	3.61	1.8
How easy was the method to learn?	5.14	1.6

on loss of money or wrong choice of goods rather than on discrete errors at particular stages of the sequence. The SHERPA method seems to encourage a greater scrutiny of the elements of the human-machine interaction and this emphasis on task decomposition can encourage the participants to focus upon isolated, discrete task elements rather than upon overall task goals. This structured approach is likely to account for the encouraging levels of reliability and validity of the method.

5. General discussion

The studies reported in this paper have sought to determine the efficacy of SHERPA as a method for human error identification as well as proposing a methodological approach for investigating human error identification techniques in general. In the introduction, four subsections were used to frame the hypotheses. These will be used again to consider the implications of the findings.

5.1. *Learnability*

Participants report that the SHERPA methodology is relatively easy to learn (see section on subjective user opinion) and this is consistent with the reports by Kirwan (1992b). The quantitative data appear to support this as participants achieved quite high levels of sensitivity after just 3 h of instruction. Over time, participants showed improvement through a reduction in the false alarm rate and a related increase in the rate of correct rejections. However, the degree of sensitivity of error prediction with SHERPA remained at a consistent and high level despite this fluctuation. These results are important because they demonstrate that the skills required for efficient and effective error prediction can be acquired and mastered within a reasonable amount of time. Given Kennedy's (1994) recommendation that HEI methods be used in conjunction with one another, it would seem important to assess the learnability of other HEI techniques. This would be valuable for two reasons. First, the documentation associated with most methods is typically poor. Readers have to infer the approach based upon papers reported in the literature. Second, successful use of HEI techniques relies upon the judgement of the analyst. Most error analysts acquire these skills over a number of years and the skills are not easy to teach to others. Studies of acquisition of these skills would lead to a call for higher quality documentation as well as providing insights into the way in which people learn how to use the approaches. This paper illustrates the form in which those studies could be conducted, i.e. studies should consider using a larger number of participants and should provide empirical evidence of reliability and validity using the signal detection paradigm. While subjective assessments of techniques provide a useful insight into the more qualititative aspects of methods, this is no substitute for quantitative data on the performance of the methods in predicting errors.

Despite the fact that SHERPA is a structured technique, there is still a good deal of reliance upon the judgement of the analyst in determining which errors are credible in any given situation. This judgement may be likened to the criterion shift in signal detection theory in determining the difference between signals (errors in this case) and noise. At least two factors play a part in the development of the sensitivity of the response operator curve: domain expertise and expertise in the human error identification method. This is an important point, because one is unlikely to yield good results without the other, as the heuristic evaluation shows. In the heuristic evaluation, the participants had a high level of domain knowledge (they were all regular users of the device under evaluation) but they had no knowledge of the human error identification technique. When domain knowledge was combined with device knowledge, a dramatic improvement in the accuracy of error identification was witnessed. As with all skills, however, improvement comes with practice (Annett 1969, Welford 1971). This may explain the anomaly apparent in the comparison between the SHERPA results at time 1 and time 3. The time gap had led to some small decrement in skill that would not have occurred with regular practice. This has important connotations for users of SHERPA who may only have cause to use it on an infrequent basis. Refresher training and prior usage after a period of absence would be recommended.

5.2. *Test-retest reliability*

Test-retest reliability of SHERPA remains fairly consistent over time, despite the variation in the delay between trials. However, the correlation coefficient for test-retest reliability was moderate (by analogy to psychometric test development) and it would have been desirable to have participants achieve higher levels of reliability. Reliability and validity are interdependent concepts, but the relationship is in one direction only. While it is perfectly possible to have a highly reliable technique with little validity, it is impossible for a technique to be highly valid with poor reliability. As Aitkin (1985) notes 'reliability is a necessary condition but not a sufficient condition for validity' (Aitkin 1985:93). Consequently, establishing the validity of a HEI technique is of paramount importance. The current investigation is the first reported study of people learning to use a HEI technique; all previously reported studies have been with expert users. As a result, it is conceivable that the moderate reliability values ($r = 0.4 - 0.6$) obtained here may simply be an artefact of lack of experience. With this in mind, it is important to note that Baber and Stanton (1996) report much higher values when users are experts.

5.3. *Concurrent validity*

The results of experiment 1 show that SHERPA provides a better means of predicting errors than an heuristic approach, and demonstrates a respectable level of concurrent validity. These findings suggest that SHERPA enables analysts to structure their judgement. However, the results run counter to the literature in some areas (such as usability evaluation), which suggest the superiority of heuristic approaches (Neilsen 1992). The views of Lansdale and Ormerod (1994:257) may help to reconcile these findings. They suggest that, to be applied successfully, an heuristic approach needs the support of an explicit methodology to 'ensure that the evaluation is structured and thorough'. In a similar vein, Kirwan's (1992b) assessment of six HEI techniques concluded that the combination of an heuristic approach with SHERPA offered the optimal means of predicting error. Essentially, SHERPA aims to do exactly what Lansdale and Ormerod (1994) prescribe. It provides a semi-structured approach that forms a framework for the judgement of the expert analyst without constraining it. It seems to succeed precisely because of its semi-structured nature, which alleviates the burden otherwise placed on the analyst's memory while allowing them room to use their own heuristic judgement. By analogy, this mirrors precisely the findings of research into the selection interview, which has shown the traditional interview approach to be both unreliable and a poor predictor of candidate performance in the job (Wagner 1949, Mayfield 1964). However, more recent research has demonstrated that the semi-structured situational interview leads to dramatic improvements in both the reliability and validity of the approach (Weisner and Cronshaw 1988). Arguably, the structure of SHERPA, with its focus on the individual tasks, make it the human error equivalent of the semi-structured situational selection interview.

5.4. *Subjective user opinion*

As in many areas, it is not sufficient to demonstrate that a method works, it must also gain acceptance from the people using it. The results of experiment 4 show that participants not only found SHERPA relatively easy to learn, but they also found it to be well structured, with an acceptably high level of face validity. These findings support previous research (Kirwan 1992b, Baber and Stanton 1996) and mean that the SHERPA methodology and its results are likely to be well accepted by users. This issue of user acceptability has been of great concern within the field of personnel selection, to the extent that some organizations have gone to considerable lengths to ensure the validity and reliability of their selection methods. Assessment centres, in which candidates participate in a range of exercises, offers one a prime example of this approach (Feltham 1989). In principle, some have envisaged a battery of HEI techniques being applied to human-machine systems (Kennedy 1994). Such an approach would certainly address one's concerns about validity and reliability of HEI outcomes; however, the cost of such a practice in terms of time and resources would make it viable only in very high-risk situations.

It has been suggested by Neilsen (1992) that one solution to the problem of variability between evaluators is to aggregate the evaluations from a number of evaluators. In a study comparing novice and expert evaluators, the numbers of problems identified increased as a function of expertise and the number of evaluators used (Neilsen 1992). Thus one would expect more human errors to be correctly predicted, with increases in both the expertise of the evaluators and the number of evaluators used. Neilsen also identified a third factor: domain expertise. He reports an interaction between domain expertise, expertise in evaluation, and number of evaluators used. These results could certainly be extrapolated to the use of HEI techniques.

6. General conclusions

The research reported in this paper offers significant advances on previous studies of human error. While previous studies were able to offer credible evaluations in terms of ecological validity, the comparisons were poor in terms of experimental rigour. Consequently, one can question whether the differences reported could be attributed to differences in the HEI techniques themselves, or due to methodological weaknesses. The present paper addresses these concerns and puts forward standard benchmarks against which all HEI techniques may be evaluated. However, one note of caution emerges from the present study. By importing the criteria from the field of psychometric testing, the authors have set themselves very high standards. Given the immeasurable complexity of human behaviour and of human error, it is possible that one is aiming too high—one is expecting HEI techniques to be able to predict the forms in which complex errors will manifest themselves armed only with a description of the task activities and an error classification system. Accordingly, by the standards of psychometric testing, one may have to adjust one's sights when evaluating HEI performance.

Nevertheless, the present evaluation of SHERPA revealed a surprisingly good performance despite the inexperience of the analysts and the simplicity of the technique. This offers an optimistic view of the future for other HEI techniques. However, these other techniques need to be evaluated comprehensively before one can be confident in recommending their use. Research should consequently be directed to three aims; first, all human error identification techniques should be evaluated in a statistically rigorous manner; second, cross-validation studies should be undertaken; and third, meta-analyses should be conducted in order to facilitate comparison of the performance of the various HEI techniques. These lines of enquiry will enable the authors to be in a stronger position when revising existing techniques and recommending their use in the field.

Concurrent validity is an obvious way of validating the SHERPA technique as the validation data are readily available. The next obvious step in the validation process is to test the predictive validity of the approach. This requires a longitudinal study to be conducted and is an important goal for future research. The signal detection paradigm provides a useful framework for testing the power of HEI techniques. In particular, it identifies type I errors (a miss: when the error analyst predicts the error will not occur and it does) and type II errors (a false alarm: when the error analyst predicts that there will be an error and there is not) in the judgement of the analyst. It would be interesting to see if other HEI techniques are more or less sensitive in structuring the judgement of the analyst.

Human error identification remains a controversial area. Some people doubt the credibility of the methods and the meaningfulness of the results (Reason 1990). Even those who think that human reliability is a worthwhile pursuit disagree on how it may best be practised. Some researchers are optimistic about decompositional approaches, such as SHERPA (Kirwan 1992b) while others argue that only higher-level analyses are meaningful (Hollnagel 1993). This debate is likely to continue for the foreseeable future. In the meantime, the authors feel that more effort should be expended on examining the existing methods in finer detail.

Acknowledgements
The authors would like to thank Mark Young and the two reviewers for their helpful comments on an earlier draft of this paper.

References
AITKIN, L. R. 1985, *Psychological Testing and Assessment* (Boston: Allyn & Bacon).

ANNETT, J. 1969, *Feedback and Human Behaviour* (Harmondsworth: Penguin).

ANNETT, J., DUNCAN, K. D., STAMMERS, R. B. and GRAY, M. J. 1971, *Task Analysis*, training information no. 6 (London: HMSO).

BABER, C. and STANTON, N. A. 1996, Human error identification techniques applied to public technology: predictions compared with observed use, *Applied Ergonomics*, **27**, 119–131.

BARTRAM, D., ANDERSON, N., KELLETT, D., LINDLEY, P. and ROBERTSON, I. 1995, *Review of Personality Assessment Instruments (Level B) for Use in Occupational Settings* (Leicester: BPS Books).

BARTRAM, D., LINDLEY, P., FOSTER, J. and MARSHALL, L. 1992, *Review of Psychometric Tests (Level A) for Assessment in Vocational Training* (Leicester: BPS Books).

BLINKHORN, S. and JOHNSON, C. 1990, The insignificance of personality testing, *Nature*, **348** 671–672.

COOK, M. 1988, *Personnel Selection and Productivity* (Chichester: Wiley).

CRONBACH, L. J. 1984, *Essentials of Psychological Testing* (New York: Harper & Row).

EMBREY, D. E. 1986, SHERPA: A systematic human error reduction and prediction approach. Paper presented at the International Meeting on Advances in Nuclear Power Systems, Knoxville, TN, April.

FELTHAM, R. T. 1989, Assessment centres, in P. Herriot (ed.), *Assessment and Selection in Organisations* (Chichester: Wiley).

HOLLNAGEL, E. 1993, *Human Reliability Analysis: Context and Control* (London: Academic Press).

KENNEDY, R. J. 1994, Can human reliability assessment (HRA) predict real accidents? A case study analysis of HRA, in A.I. Glendon and N.A. Stanton (eds), *Proceedings of the Risk Assessment and Risk Reduction Conference*, Aston University, Birmingham, March (Birmingham: Aston University).

KIRWAN, B. 1992a, Human error identification in human reliability assessment. Part 1: Overview of approaches, *Applied Ergonomics*, **23**, 299–318.

KIRWAN, B. 1992b, Human error identification in human reliability assessment. Part 2: Detailed comparison of techniques, *Applied Ergonomics*, **23**, 371–381.

KIRWAN, B. 1994, *A Practical Guide to Human Reliability Assessment* (London: Taylor & Francis).

KIRWAN, B. and AINSWORTH, L. K. 1992, *A Guide to Task Analysis* (London: Taylor & Francis).

LANSDALE, M. W. and ORMEROD, T. C. 1994, *Understanding Interfaces* (London: Academic Press).

MACMILLAN, N. A. and CREELMAN, C. D. 1991, *Detection Theory: A User's Guide* (Cambridge: Cambridge University Press).

MAYFIELD, E. C. 1964, The selection interview: a re-evaluation of published research, *Personnel Psychology*, **33**, 239–260.

NEILSEN, J. 1992, Finding usability problems through heuristic evaluations, *Proceedings of the ACM Conference on Human Factors in Computing Systems* (Amsterdam: ACM Press).

NORMAN, D. A. 1988, *The Psychology of Everyday Things* (New York: Basic Books).

PATRICK, J. 1992, *Training: Research and Practice* (London: Academic Press).

REASON, J. 1990, *Human Error* (Cambridge: Cambridge University Press).

SENDERS, J. W. and MORAY, N. P. 1991, *Human Error* (Hillsdale, NJ: Lawrence Erlbaum).

SHACKEL, B. 1990, Human factors and usability, in J. Preece and L. Keller (eds), *Human-computer Interaction* (Hemel Hempstead: Prentice-Hall).

SHEPHERD, A. 1989, Analysis and training in information technology tasks, in D. Diaper (ed.), *Task Analysis for Human Computer Interaction* (Chichester: Ellis Horwood).

THIMBLEBY, H. 1991, Can humans think? *Ergonomics*, **34**, 1269–1287.

WAGNER, R. 1949, The employment interview: a critical review, *Personnel Psychology*, **2**, 17–46.

WEISNER, W. H. and CRONSHAW, S. F. 1988, A meta-analytic investigation of the impact of interview format and degree of structure on the validity of the employment interview, *Journal of Occupational Psychology*, **61**, 275–290.

WELFORD, A. T. 1971, *Fundamentals of Skill* (London: Methuen).

WHALLEY, S. J. and KIRWAN, B. 1989, An evaluation of five human error identification techniques. Paper presented at the 5th International Loss Prevention Symposium, Oslo, June.

Wickens, C. D. 1992, *Engineering Psychology and Human Performance* (New York: Harper Collins).

Williams, J. C. 1989, Validation of human reliability assessment techniques, *Reliability Engineering*, **11**, 149–162.

Appendix. Frequency of prediction of each error type using SHERPA†

Error type	Frequency at time 1	Frequency at time 2	Frequency at time 3
Money at wrong angle	21	15	10
Put wrong coins in	12	7	6
Fail to put coins in	25	23	21
Not enough money in	12	11	18
Pressed wrong character	20	16	18
Fail to press character	20	21	21
Pressed wrong number	21	17	17
Push flap too early	20	21	19
Insufficient turn of handle	12	11	14
Fail to turn handle at all	24	23	20
Fail to pick up change	23	24	23

†Based on responses of a set of 25 participants taking part at all three testing phases

CHAPTER 13

Future directions for task analysis

NEVILLE A. STANTON and JOHN ANNETT

1. Introduction

There has been considerable development in the field of task analysis since the days of Taylor and Gilbreth, as shown in the review of the field and further illustrated by the contributions throughout this text and in other books (see table 1 for a summary of the number of methods available). The problem with tables like this is that they will always be incomplete. (Apologies to anyone who feels left out!) There are some worthwhile texts that do not fit neatly into our analysis (such as Essens *et al.*'s (1994) book that mentions many methods and distinguishes between task analysis, performance analysis, cognitive task analysis and cognitive performance analysis) and texts that are just about to be published in this growing field (such as Schraagen *et al.*'s (1999) book on cognitive tasks analysis based on a recent workshop on methods in current use). As a concluding chapter, we speculate about what further developments are likely to occur and propose future directions for research and development effort. In order to conduct this exercise, we have framed the future in terms of theoretical progress, development of methods and mechanisms for conducting analysis. Task analysis, perhaps more than any other aspect of ergonomics, has explicit links between theories of human performance and methods for analysing human activity. We believe this to be a push-pull relationship, such that theories of human activity encourage the development of methods of analysis (theory-driven push) and that the analysis of human activity informs the development of theory (the analysis-driven pull). Both forms of analysis are necessary to stimulate progress. Given that task analysis is one of the fundamental cornerstones of ergonomics, much rests upon progress in this area.

2. Theoretical progress

Theoretical progress is likely to be made through the interweaving of several current strands in the pure and applied sciences. From the discipline of psychology, we need further progress in the understanding of human cognition and behaviour in individuals and groups, for example how stress affects higher cognitive functions. From ergonomics, we need further development of the systems paradigm and in the understanding of human ecology on behaviour. The systems view has offered an enduring framework for task analysis, but various techniques have tended to concentrate only on a limited sub-set of the possible interactions, e.g. the tasks conducted by an individual, or the use of a device, or the communications between individuals. No single technique offers a comprehensive account of human activity that can account for all manner of interactions (i.e. from devices, to environment, to people, to teams), with differing levels of skill (i.e. from novice to expert), with reference to both cognitive and physical demands. While this might seem a tall order for task analysis, it must remain a goal. One of the main task analysis techniques, HTA, does come close in many respects. Back in the early 1970s (Annett *et al.* 1971) wrote: 'The process of analysing a task . . . is the process of diagnosing the plan which is needed to achieve a stated goal.'

Table 1. Other texts with the number of tasks analysis related methods.

Author(s)	Title	Edited/ Authored	Date (ed)	Pages	Number of methods
Diaper	*Task Analysis in Human Computer Interaction*	Edited	1989 (1st)	258	6
Kirwan and Ainsworth	*A Guide to Task Analysis*	Edited	1992 (1st)	417	23
Kirwan	*A Guide to Practical Human Reliability Assessment*	Authored	1994	592	28
Corlett and Clarke	*Ergonomics of Workspace and Machines*	Edited	1995 (2nd)	128	6
Wilson and Corlett	*Evaluation of Human Work*	Edited	1995 (2nd)	1134	48
Jordan *et al.*	*Usability Evaluation in Industry*	Edited	1996 (1st)	252	20
Salvendy	*Handbook of Human Factors and Ergonomics*	Edited	1997 (2nd)	2137	over 100
Seamster *et al.*	*Applied Cognitive Task Analysis*	Authored	1997 (1st)	338	8
Stanton	*Human Factors in Consumer Products*	Edited	1998 (1st)	287	27
Stanton and Young	*A Guide to Methodology in Ergonomics*	Authored	1999 (1st)	132	12

What they had in mind, and this is illustrated in many of the analyses later carried out by Keith Duncan and Andy Shepherd, was that HTA would identify what the operator had to know. For example, the meaning of various possible stimulus patterns (both input and feedback) and rules and principles which determine (or inform) decisions and actions. One of the main purposes of the people involved in the development of HTA was to add a cognitive dimension to the (then) standard methods of analysing tasks. Clearly they anticipated CTA by a decade or more!

3. Development of methods

Attempts to tie the various theoretical and methodological strands together will only come about by framing the problem in terms of human activity in context. This ecological perspective seems to be fundamental to task analysis. Some progress has already been made in this direction, which has been achieved either by extending current approaches or developing time-line techniques that trace activity and inter-actions over a sample period. Embracing context in this way seems to be an increasingly popular method for analysing human activity, and suggests that behaviour results as much from the environment within which it is performed as it does the goals of the actor within it.

Many approaches developed upon contextual analysis are to be found within this book, for example the Pentanalysis proposed by Diaper *et al.* (a human-machine interaction technique), the CUD method put forward by Watts and Monk (a multi-agent time-line approach), ACTA method proposed by Militello and Hutton (a method for determining knowledge requirements), the SGT technique developed by Ormerod *et al.* (a structured approach based on HTA), TTRAM proposed by Swezey *et al.* (a training requirements methodology), case studies of CTA by O'Hare *et al.* (based on training and interface design examples), and the SHERPA method discussed by Stanton and Stevenage (an error analysis approach). These examples show that a greater understanding of the consequences of behaviour can be appreciated through a contextual analysis, such as the interactions between people and technology related to the task in CUD and the potential for human error impact on the system in SHERPA.

There is a concern within ergonomics that many of the methods developed are only ever used by their developers and have little significance for others. Hopefully the reader will be convinced of the usefulness of the approaches presented in this book, all of which have undergone extensive peer review. Further developments are likely to occur through consensus on the theoretical underpinning and agreement upon the usefulness of the methods. What we are likely to see is convergence between all of the methods, leaving a core of methods which satisfy most of the needs most of the time (see Shryane *et al.* in this book). Thus, we predict theoretical development will lead to fewer, rather than more, methods. The general proliferation of methods at present is an artefact of conceptual stumbling blocks that prevent unification. This presents the problem of having very specialized methods for examining specific aspects of human activity. For example, different methods exist for analysing performance time, action, knowledge requirements, skills, errors, communication, gaze shifts between devices, movement, reach, etc. The combination of all of these aspects of performance into a single framework would certainly be welcome, if the resultant analysis could offer reliability and validity. At present, little is known of the reliability and validity of ergonomics methods, although some studies are starting to be presented in the open literature (Stanton and Young 1999). There is certainly plenty of scope for more research of this kind to be conducted, and the development of methods would clearly be beneficial. In addition, standardized training and documentation of methods should be developed, as acquisition appears to rely largely upon self-development at present.

Perhaps the reasons that HTA has achieved such longevity (approaching three decades of sustained use) is that it was developed upon a theory of human performance and it has application to many potential uses (such as training, personnel selection, interface design and human reliability), and a formal training in the technique may be found as part of a Masters programme or from a consultancy company. In addition, there are an increasing number of packages offering computer-based support for HTA. Typically, these have been developed by ergonomics consultancies for their own use that have subsequently been made available for others to purchase, as either part of a training programme or separately. This serves as a excellent template for further development of methods.

4. Mechanisms for conducting task analysis

The main benefit of task analysis techniques, when performed correctly, is that they structure information about the task. This requires that the analyst is sufficiently competent in their ability to sift, classify, and determine the appropriateness of the information and overlay the structure on to it. The inability to perform these operations correctly is all too clear in the novice analyst, when the information is typically presented in a linear, procedural, manner. This can be corrected through training and practice together with feedback from the trainer. However, some (e.g. Dowell and Long, 1998) have suggested that the implicit knowledge requirement to structure the analysis is a weakness of the methods, rather than of the analyst. Some (e.g. Shepherd in this book) have attempted to correct this position through the development of more formal approaches to task analysis. The formality of such methods is something of an illusion, however. Just because a method is more mechanistic than another, this does not mean that the analysis is any more correct. Quite the reverse could be true if the mechanisms were so restrictive as to force unnatural descriptions of tasks from the analyst. In R. B. Miller's (1953) report on A Method of Man-Machine Task Analysis he presents his approach as a set of guidelines rather than as a rigidly prescriptive method. This inherent flexibility may, in the long run, have rather more success than approaches with too rigid a structure. Others have approached the problem by developing techniques to support the analysis of tasks through existing methods (e.g. the software to support task taxonomies and plans by Ormerod *et al.* in this book) and the development of meta-methodologies that join existing methodologies together (e.g. the Pentanalysis technique by Diaper *et al.* in this book). Both of these approaches share the characteristic of overlaying a structure on the process of analysing tasks, without being overly mechanistic. The authors of these approaches argue that structuring the methods in this way does not interfere with analysis; rather benefit is to be gained by making the analysis rather more consistent, as well as assisting novice analysts in the first instance.

Mechanisms for collecting the raw data could similarly be structured. Task analysis, of existing tasks, tends to rely largely upon observation, interviews and documentation. While techniques for analysing these data independently are available, these have yet to be integrated and combined with the task analytic method. Conceptually, at least, this is not a difficult undertaking, but one that has yet to be proven in the field.

5. Conclusions

While much research and development effort has been expended upon task analysis, a fact to which the contributions to this book attest, there is still further distance to travel. It is our assertion that the effort needs combined theoretical and methodological progress. While we can be pleased with the results of effort to date, and the enduring nature of HTA, this should give no room for complacency. The centrality of task analysis to a range of applications, as shown throughout this book is illustrated in figure 1.

The hub of the research revolves around the development of adequate tools and methods of tasks analysis. Ainsworth and Marshall have reviewed the state of the art in a variety of domains, while Shepherd focuses on the mainstay technique of HTA. Ormerod *et al.* consider the use of computerized support to HTA as a means of supplementing expertise. Application of task analysis techniques to the assessment of multiple-person scenarios was undertaken by both Marti and Watts and Monk. As team-based work is an essential feature of most work places, it is important to be able to analyse co-operating groups as well as individuals. Militello and Hutton concentrated on the analysis of knowledge requirements and the assessment of training needs. Training needs analysis also featured in the work of Swezey *et al.* and

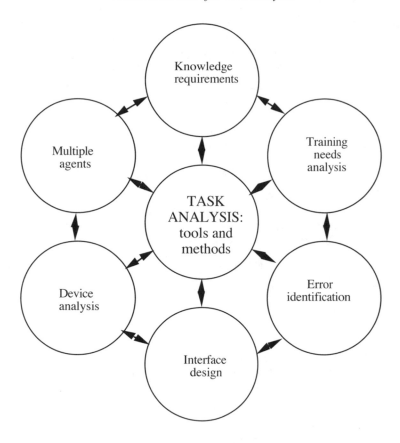

Figure 1.   Applications of task analysis.

O'Hare *et al.*, the latter of which also considered the implications for interface design. Human error identification was undertaken by Stanton and Stevenage and by Shryane *et al.* Finally, Diaper et al sought to examine the relations between artefacts and humans in a given task scenario. All of the research papers had some form of task analysis and, as is indicated in figure 1, there are links between the areas of application. For example, error identification can lead to the proposal for training needs and recommendations for changes in interface design. Similarly, analysis of multiple agents can lead to the identification of knowledge requirements and deepen the understanding of how devices are used. At the core of all these applications is task analysis, which can mediate the links between applications. We cannot overstate the importance of the role played by task analysis, which justifies the considerable research and development effort expended in this endeavour.

Some of the issues to emerge from the discussions by the contributors include:

- How deep should the analysis be?
- Which methods of data collection should be used?
- How should the analysis be presented?
- Where is the use of the method appropriate?
- How much time/effort does each method require?

- How much, and what type, of expertise is needed to use the method(s)?
- What tools are there to support the use of the method(s)?
- How reliable and valid is/are the method(s)?

Answering these questions is likely to keep the ergonomics research community busy for the foreseeable future

References

ANNETT, J., DUNCAN, K. D., STAMMERS, R. B. and GRAY, M. J. 1971, *Task Analysis* (HMSO, London).

CORLETT, E. N. and CLARKE, T. S. 1995, *The Ergonomics of Workspaces and Machines*, 2nd edn. (Taylor & Francis, London).

DIAPER, D. 1989, *Task Analysis in Human Computer Interaction* (Ellis Horwood, Chichester).

DOWELL, J. and LONG, J. 1998, Conception of the Cognitive Engineering Design Problem, *Ergonomics*, **41**, 126–139.

ESSENS, P. J. M. D., FALLESEN, J.-J., McCANN, C. A., CANNON-BOWERS, J. and DORFELL, G. 1994, COADE—A Framework for Cognitive Analysis, Design and Evaluation. NATO Defence Research Group RSG 19. TNO, Soesterberg.

JORDAN, P. W., THOMAS, B., WEERDMEESTER, B. A. and McCLELLAND, I. L. 1996, *Usability Evaluation in Industry* (Taylor & Francis, London).

KIRWAN, B. 1994, *A Guide to Practical Human Reliability Assessment* (Taylor & Francis, London).

KIRWAN, B. and AINSWORTH, L. 1992, *A Guide to Task Analysis* (Taylor & Francis, London).

MILLER, R. B. 1953, *A Method for Man-Machine Task Analysis*, Wright-Patterson Air Force Base, Ohio. Wright Air Development Center Technical Depot, 53–137.

SALVENDY, G. 1997, *Handbook of Human Factors and Ergonomics*, 2nd edn. (Wiley, New York).

SCHRAAGEN, J.-M., CHIPMAN, S. and SHALIN, V. 1999, *Cognitive Task Analysis* (Erlbaum Hillsdale, NJ,).

SEAMSTER, T .L., REDDING, R. E. and KAEMPF, G. L. 1997, *Applied Cognitive Task Analysis in Aviation* (Avebury, Aldershot).

STANTON, N. A. 1998, *Human Factors in Consumer Product Design* (Taylor & Francis, London).

STANTON, N. A. and YOUNG, M. S. 1999, *A Guide to Methodology in Ergonomics* (Taylor & Francis, London).

WILSON, J. R. and CORLETT, E. N. (1995) *Evaluation of Human Work*, 2nd edn. (Taylor & Francis, London).

GLOSSARY

ACT*	Adaptive Control of Thought
ACTA	Applied Cognitive Task Analysis
ANOVA	Analysis of Variance
ANSI	American National Standards Institute
ATC	Air Traffic Control
ATHOS	Airport Tower Harmonized Controller System
BARS	Behaviourally Anchored Rating Scales
BNF	Bachus - Naur Form
C	Consultant
CASE	Computer Assisted Software Engineering
CCI	Critical Cue Inventories
CDM	Critical Decision Method
CGA	Conceptual Graph Analysis
CSCW	Computer Supported Co-operative Work
CTA	Cognitive Task Analysis
CTM	Composite Task Model
CUD	Communications technologies Usage Diagram
CWP	Controller Working Positions
DDE	Data Dictionary Entries
DFD	Data Flow Diagram
DIS	Distributed Interactive Simulation
EW	Electronic Warfare
GC	Ground Controller
GDSS	Group Decision Support Systems
GIS	Graphic Information System
GOMS	Goals, Operators, Methods and Selection rules
GP	General Practitioner
GTM	Generalized Task Model
HAZOP	Hazard and Operability study
HCI	Human Computer Interaction
HEI	Human Error Identification
HEP	Human Error Probabilities
HTA	Hierarchical Task Analysis
IAF	Input - Action - Feedback
ISO	International Standards Organization
IWS	Interactive Work System
JSD	Jackson System Design
LASCAD	London Ambulance Computer-Aided Dispatch project
M	Mean
MDS	Multidimensional Scaling
MSDS	Material Safety Data Sheets
MTC	Minor Treatment Centre
MUSE	Method for Usability Engineering
NDM	Naturalistic Decision Making

NP	Nurse Practitioner
P	Patient
PARI	Precursor, Action, Result and Interpretation
PC	Personal Computer
PHECA	Potential Human Error Causes Analysis
PRA	Probabilistic Risk Assessment
R	Relative
RPD	Recognition-Primed Decision
SAR	Situation Assessment Record
SD	Standard Deviation
SGT	Sub-goal Templates
SHERPA	Systematic Human Error Reduction and Prediction Approach
SME	Subject Matter Experts
SMGCS	Surface Movement Ground Control System
SRK	Skill-Rules-Knowledge
SSADM	Structured System Analysis and Design Method
SSI	Solid State Interlocking
TA	Task Analysis
TAG	Task Action Grammar
TAKD	Task Analysis for Knowledge Descriptions
TC	Tower Controller
THERP	Technique for Human Error Rate Prediction
TMC	Teleconferenced Medical Consultations Project
TOTE	Test-Operate-Test-Exit
TTRAM	Task and Training Requirements Analysis Methodology
UK	United Kingdom
USA	United States of America
VDU	Visual Display Unit
VFR	Visual Flight Rules
WISSARD	What-If Simulation System for Advanced Research and Development

Author index

Subject index